Build Your Own Home from St

Go Through a Step-by-Step Process of:
-       **Determining House Specs**
-       **Finding an Owner / Builder Program**
-       **Selecting a Lot**
-       **Getting a Contractor Loan**
-       **Selecting Sub-Contractors**
-       **Building Your House**

And Save as much as 25% of the Total Construction Costs

Build Your Own Home from Start to Finish

Go Through a Step-by-Step Process of:
- **Determining House Specs**
- **Finding an Owner / Builder Program**
- **Selecting a Lot**
- **Getting a Contractor Loan**
- **Selecting Sub-Contractors**
- **Building Your House**

And Save as much as 25% of the Total Construction Costs

by Dennis and Nancy Sue Swoger

QP Publishing, Finleyville, Pennsylvania

Build Your Own Home from Start to Finish

Go Through a Step-by-Step Process of:
- **Determining House Specs**
- **Finding an Owner / Builder Program**
- **Selecting a Lot**
- **Getting a Contractor Loan**
- **Selecting Sub-Contractors**
- **Building Your House**

And Save as much as 25% of the Total Construction Costs

by Dennis and Nancy Sue Swoger

Published by:

QP Publishing
PO Box 237
Finleyville, Pennsylvania 15332-0237 USA

Copyright © 1999 by Dennis and Nancy Sue Swoger
Printed in the United States of America

Morris Publishing
3212 E. Hwy 30 • Kearney, NE 68847•
1-800-650-7888

Library of Congress Cataloging in Publication Data - Card Number Preassigned is: 99-093232

Swoger, Dennis and Nancy Sue
Build Your Own Home from Start to Finish: Go Through a Step-by-Step Process of Determining House Specs, Finding an Owner / Builder Program, Selecting a Lot, Getting a Contractor Loan, Selecting Sub-Contractors, Building Your House and Save as much as 25% of the Total Construction Costs / by Dennis and Nancy Sue Swoger

CIP 99-093232
ISBN 0-9626692-9-6

We dedicate this, our first joint book,
to our great Lord
who has given us
wisdom through prayer
to accomplish this home building project.

Table of Contents

Warning - Disclaimer

INTRODUCTION

Welcome! This Workbook came about after many fun, frustrating, loving moments we have had in building our home through an Owner-Builder Program. We want to share our experiences and gained knowledge with you - that you may bypass the pitfalls we encountered (and believe me, we encountered a host of them!).

Let's begin by first telling you what it means to be an "Owner-Builder" and why we started down this path of Owner-Builder. Then we will get into the "How-to" details of how you can do it yourself, save lots of money or afford a house that is larger, more luxurious than what you first thought you could afford. Be your own General Contractor and call all the shots!

An Owner-Builder is a term used to describe a couple (or individual) who elect to build a home by themselves using plans and possibly materials purchased from a company who sells home packages. You will act as the General Contractor (unless you opt to hire someone else to do this job). The significant savings in building comes when you act as your own General Contractor.

Throughout this book we will include some tips that you will not have learned without experience. We learned early in the game that the Owner-Builder Programs give you just (barely) enough to get by and all the rest comes with experience. We want to increase your chances of Success and decrease your out-of-pocket expenses. Some of these "Hot Tips" will be highlighted in a Tip Section like the Tip on Page 3.

We have also included plenty of space for Notes. You will want to make sure you keep notes of things you want or need to do along the way. You may also want to keep a Home Building Diary of what was done and when it was done. Things you may want to include in a diary are: Contractor Information (date promised, date started, date finished, missed days); Delays (weather, supply delays, Contractor delays); Materials uses (manufacture, model, color); and, Appliance Information (date purchased, date installed, date began to use, types of warranties).

Tip

Expect this project to take much longer than is predicted by the Owner-Builder Representative. We found that the Sub-Contractors tend to place your job at a lower priority than other jobs with Professional Contractors. They look at your project as a one time job. Most Sub-Contractors work for various General Contractors throughout the building season and will cater to their needs before yours.

Now let's begin!

For which one of you, when he wants to build a tower, does not first sit down and calculate the cost, to see if he has enough to complete it? Otherwise, when he has laid a foundation, and is not able to finish, all who observe it begin to ridicule him, saying "This man began to build and was not able to finish". - Luke 14:28-30, NAS Study Bible

CHAPTER ONE

Finding the Home that Fits You

Before we knew that we would build our "Dream Home" ourselves, we visited so many existing homes. There were so many variables we were considering including:

Style
> We wanted a modern, single unit home.

Number of Stories
> We were looking for 2-Stories. However, in our travels we were pleasantly surprised to see how large Ranch-Style homes have become.

Number of Bedrooms
> We wanted 4, but 3 or 5 could be considered.

Kitchen Size
> We enjoy cooking together and desired lots of counter space and room to move around.

Master Bedroom Size
> We wanted one at least 18' by 12'. This is what we had in the townhouse we currently owned.

Garage
> Minimum of a two car garage required.

Age
> We didn't want to look at a home much older than 10 years. Newer homes offer modern appliances, whirlpool tubs, and still room to negotiate the price.

Variables considered when looking for our home - continued

Covenants

Living in a townhouse lent itself to being swamped with restrictive covenants. We were ready for a little more freedom.

Taxes

Moving to another County would drop our property tax milage by a third. Unfortunately, people selling homes in these areas also knew this and were raising their asking prices.

After looking at hundreds of used and new homes, we "stumbled" across a Model Home of a company who built homes, but also offered the "Owner - Builder" Program. The Model Home that we viewed looked "perfect". The price tag of the Model Home was steep, which prompted us to investigate what it would cost if we built it ourselves.

This was the beginning of our Owner-Builder Relationship.

Remember, when you see something you like, make a quick sketch or get it on film or videotape. Cut pictures of items you like from magazines or newspapers. By accumulating these ideas, you can focus on the look you like. As you change your mind, you can replace the items with new stuff. Log your ideas and wants on the **Desired Home Features Checklist** which spans the next few pages. This will come in handy as you work through the process that everything is a decision!

How lovely are Thy dwelling places, O Lord of hosts! - Psalm 84:1, NAS Study Bible

Desired Home Features Checklist

Item	Must Have	Nice to Have
- Whole House Items -		
Alarm System		
Central Vacuum System		
Central Air Conditioning System		
Ceiling Fans - Locations ____		
Ceiling Fans - Number ___		
Intercom System		
Cable Service		
Phone Service		
Phone Lines - Number		
Heating Style (electric, gas)		
Fireplaces - Number ___		
Fireplaces - Gas		
Fireplaces - Log Burning		
Fire Sprinkler System		
Flooring - Carpeting - Wall to Wall		
Flooring - Hardwood		
Flooring - Marble		
Flooring - Ceramic		
Lighting - Recessed		
Satellite Dish		
Size - Minimum Square Footage _____		
Skylights - Number _____		
Skylights - Location _____		
Sound System		
TV Cable		
Wall Covering - Paneling		
Wall Covering - Wall Paper		
Wall Covering - Paint		
Water Supply - Public		

Desired Home Features Checklist

Item	Must Have	Nice to Have
- Whole House Items - continued -		
Window Coverings		
Window Sills		
Windows - Number & Style		
- Attic and Basement -		
Attic - Finished		
Skylights in the Attic		
Basement - Unfinished		
Basement - Finished		
Bathroom in the Basement		
Access door from the Basement to Outside		
Windows in the Basement - Number ___		
Laundry Area in the Basement (?)		
- Bathrooms -		
Bathrooms - Number __		
Bathroom attached to each Bedroom (?)		
Skylights in Bathroom(s)		
Ceiling Fans in Bathroom(s)		
Cathedral Style Ceiling in Bathroom(s)		
Counter Top Material		
Cabinet Material and Style		
Master Bath - Fireplace		
Master Bath - Whirlpool Tub		
Master Bath - Separate Counters		
Master Bath - Number of Sinks ___		
Master Bath - Flooring Material		
Master Bath - Heated Flooring		
Master Bath Size ___ and Layout		
Master Bath - Type of Shower (or Tub / Shower)		
Number of Sinks in other Bathrooms ____		

Desired Home Features Checklist

Item	Must Have	Nice to Have
- Bedrooms -		
Bedrooms - Number ___		
Ceiling Style - Cathedral		
Ceiling Style - Tray		
Ceiling Style - _____ (other)		
Closets - Walk-in		
Closets - Number in Master Bedroom _____		
Master Bedroom Size _____		
Master Bedroom Fireplace		
Master Bedroom - Multiple Cable Outlets		
-Kitchen -		
Breakfast Nook Area		
Cabinets - Style		
Cabinets - Material / Wood Type		
Cabinets - Color		
Counter Tops - Tile		
Counter Tops - Other		
Counter Tops - Granite		
Counter Tops - Marble		
Dishwasher		
Eat-In Kitchen		
Garbage Disposal		
Microwave Oven Built-In		
Morning Room		
Oven / Stove - Electric		
Oven - Built-In		
Oven / Stove - Gas		
Refrigerator - Style ____		
Stove Top - Drop In		
Stove / Oven (1 Piece)		

Desired Home Features Checklist

Item	Must Have	Nice to Have
- Den / Office -		
Den / Office (yes or no?)		
Den / Office - Door(s) to Enter		
Den / Office - Built in Bookshelves		
Bay Style Windows		
Separate Computer Phone Line		
Separate Fax Phone Line		
- Foyer -		
Foyer - Door Style (single or double doors)		
Foyer - 2 Story		
Foyer - Single Story		
- Other Rooms -		
Exercise Room		
Great Room		
Loft		
Sauna		
Sun Room		
Workshop		
- Garage -		
Garage - Attached (Size ____)		
Garage - Integral (Size ____)		
Garage Door Openers		
Service Door from Garage to Outside		
Height of Garage Doors		
Number of Garage Doors (doubles or singles)		
Windows in the Garage (Number ___)		
- Laundry Room -		
Laundry Room Location ____		
Cabinets in the Laundry Room		
Washtub - Number		

Desired Home Features Checklist

Item	Must Have	Nice to Have
- Laundry Room - continued -		
Counter Space		
Ironing Facility		
Cable Outlet		
Phone		
- Outside the House -		
Deck		
Driveway - Blacktop		
Driveway - Cement		
Exterior - Brick		
Exterior - Siding		
Exterior - _____ (other)		
Fenced Yard		
Hot Tub		
Landscaped - Professionally		
Patio		
Pool - in ground		
Pool - above ground		
- Your Own Items to Include -		

Who Sells Owner / Builder Homes

We live in Southwestern Pennsylvania. A well-known home construction company in this area who offered the Owner-Builder Program made this task of finding a company easy. Seeing a model home was a blessing too. We frequently visited the model (it was approximately 10 miles away from our home site) for measurement checks, placement of items, and even a photo shoot of the entire model (photographs and video tape). We have frequently referred to these photos, including taking them to stores for purchases and showing to Sub-Contractors for estimates and work questions on items to be completed.

Places for you to find Owner-Builder Programs:

- Homes For Sale section in your local newspaper

- Local Home Buying Flyers / Magazine

- Home Sales on the Internet

- Home Shows

- Local Real Estate Offices

The rain came down, the streams rose, and the winds blew and beat against that house; yet it did not fall, because it had its foundation on the rock. - Matthew 7:25 NIV

Tip

If you are blessed in having an existing model home of your plan available, be sure to thoroughly photograph and video tape all the details. You never know when the model may get sold.

Also, remember to check the size of the furniture used if the model home is decorated. We were "fooled" with a few rooms in thinking they were larger than they really were. The "trick" occurred with the use of smaller than typical furniture placed in the small rooms.

Peace and prosperity to you, your family, and everything you own! - 1Samuel 25:6 - NLT Study Bible

CHAPTER TWO

Finding the Right Property

Now that you have an idea of the house you think you want to build, you have to find where you want to build this dream home. There are things to check when selecting your property site.

When determining the size of the property, remember to consider how you will lay out the house for all the activities that will be going on during this process. Determine where you will secure the numerous building supplies that will be arriving throughout the process (framing wood, bricks, stone, sand, windows, beams, doors, siding). Have a location to store the trash bin(s) for easy access to the trash "generators" and the trash "haulers". And make sure you do not forget to obtain a portable toilet facility. We located one close to the road and away from the main action.

If building is occurring find out if anyone has had problems with excavating their property. Will rock be a problem if your land will require digging for a basement? Has the property been previously filled and if so, how long ago? How will the weather effect your property (water damage, accessibility in the winter, creek overflow)?

What is in the immediate surrounding community that could effect the smell or sound you experience (railroad tracks; community park; manufacturing or waste site; super highway)?

Flip this page to find a **Things to Check before Purchasing Property** checklist to review when looking for the right property for your home site.

Things to Check before Purchasing Property...

Item	Findings
Acreage (size)	
Building Restrictions	
Contractor Restrictions	
Corner Lot	
Covenant Restricted Property	
Easements - Any?	
Environmental Conditions (local waste sites)	
Flood Hazard Area	
Location - Close to Amenities	
Location - Rural / Country	
Location - Suburban	
Location - Urban / City	
Lot - Level	
Lot Drainage	
Lot - Clear of Woods / Forest	
Lot - Shape	
Mining Report (distance of mines below surface)	
Property Age of Neighborhood (min - avg - max)	
Property Values of Neighborhood (min - avg - max)	
School District	
Street Accessibility	
Street Lights	
Street Surface	
Tax Millage	
Township Restrictions Acceptable	
Utilities in place (city water, sewage, storm drainage)	
View (direction of weather and sun in the morning and evening)	
Walk distance to Schools	
Walk distance to Downtown	
Zoning Classification - Single Family	

Lot Size and Location

Review the house plan layout, including any attached or side entry garage space needed, with the land plot plan. Make sure your review includes the minimum set-back for the front (Building Line) and sides of the property. Our property required a 35' set-back from the front and 20' in from one side (we could choose). This made finding a lot wide enough to accept an attached, side-entry garage part of our "must have" requirements. Also, now that we are in our home we realize we could use more property. Remember when purchasing your lot, it's easier to landscape or cover property you already own (if you buy a larger lot than you thought you needed) than extend property (purchase adjoining property) after you have made your purchase. It's easier to make a better deal on a larger plot of land during the initial purchase than trying to negotiate for an adjoining lot that the seller knows you want due to your existing lot location.

Obtain a street map from the local township / borough office. This can be used to identify any possible undesirable conditions (railroad, dump site, manufacturing site, river or creek nearby) or desirable conditions (close to amenities, near a park or school). You should also drive by the community during the week-day, evening and week-end, listening for loud noises and smelling for foul odors.

If the School District is important, check the area for the state rating of the School District that you may be selecting. Often, home re-sale is largely determined by School District so don't quickly overlook this item. The School District also will make up a portion of your home owner taxes (School District tax millage). We found this to typically be the largest portion of your property taxes.

Will your property be close to a fire hydrant, street light, busy traffic? Is the lot part of a massive home plan where you will have most of the traffic passing by your home? Is the lot in a position where it will be difficult to exit from your driveway?

Each location offers its own charm. The City offers a fast-moving environment; night life and entertainment; less feelings of isolation; hospitals nearby; good Police and Fire protection; close to schools and churches; close to shopping; convenient utilities and public services; rapid and inexpensive public transportation; and, libraries and museums. The Suburbs offer the securing of having neighbors near without the crowding; a quieter place than the city; stable and appreciating property values; less crime; cleaner air; more privacy; larger lots; and, open playgrounds and parks. And the country offers lots of isolation and quiet; typically low real estate taxes; clean air; not much crime; freedom from prying eyes of neighbors; and, plenty of property for farming and raising gardens. Keep in mind that country living may also require longer drives to schools, churches, recreation, work and shopping; private water and septic systems; and there may not be many children nearby for young ones to play with.

The curse of the Lord is on the house of the wicked, but He blesses the dwelling of the righteous - Proverbs 3:33, NAS Study Bible

---

Tip

Take the time to walk the parcel of land with the Seller, looking at what you will be encountering on all sides of the property. Ask if there are any easements on the property. Make sure you are getting what you expect to be getting.

Restrictive Covenants and / or Homeowners' Associations

We have lived in communities where there were restrictive covenants and we have seen places where none existed. It was important to us that there were sensible covenants, but not too restrictive. We want to maintain the community image, thus protect the value of our property, but at the same time, we do not want to seek approval for every exterior decision. Keep this in mind when deciding whether you want covenants or not.

Some examples of items included in the covenants:

- all buildings must meet a certain minimum square footage

- all buildings are for single family residential purposes exclusively

- all homes shall use a minimum amount of brick or stone as specified

- all front porches and walks shall be concrete or a concrete product

- all homes must have a minimum number of shrubs and trees as specified

- fencing limited to backyard and only of certain permitted materials

- no parking of commercial or recreational vehicles outside the home

- no signs of any character (except "for sale") are permitted on the premises

- no outside antennas

- no outside kennels or dog houses permitted

Check to see if all outside work (driveway, landscaping, shrubs, lights, painting, sidewalks, porches, ...) must be completed before occupancy. You will want to make sure you budget for this if you are planning to finish the house after you move in.

If the property location you are considering has covenants, make sure you get a copy before you make any decision about the property. You may be buying into something that does not fit your lifestyle. Community covenants are permitted to be more restrictive than your local borough or township. They cannot be in conflict with the local government rules. Typically you are charged a monthly fee as part of the "Association" agreement. Find out what services your monthly fee will cover and what the maximum allowable increase can be per year.

Examples of services that your fee may cover:

- outside building maintenance

- snow removal

- lawn care

- fees for managing the Association

- taxes for common properties

- pool or clubhouse upkeep

- legal fees

Note: The maximum allowable increase listed in the covenants where we lived was 5% per year plus any special assessment. The "Board" running the Association always issued the maximum increase (5%) for our monthly fee. This can add up to a hefty sum over the years.

There may be restrictions as to if you can build the home yourself or if you are required to hire a Contractor from a given list determined by the Association. If you plan on being your own Contractor, you will want to make sure you discuss this with the seller / Association before exchanging any money. There may also be a restriction concerning the time you must start your construction from the date of purchase, and allowable time to complete.

Also obtain a copy of the township ordinances to review all the restrictions for the parcel of land you are interested in purchasing. It's better to do all your homework up front than pay later in the form of having the home you did not dream of or spending your hard-earned money to resolve issues.

Is it right to pay taxes to Caesar or not? But Jesus, knowing their evil intent, said "You hypocrites, why are you trying to trap me? Show me the coin used for paying the tax." They brought him a denarius, and He asked them, "Whose portrait is this? And whose inscription?" "Caesar's," they replied. Then He said to them, "Give to Caesar what is Caesar's, and give to God what is God's." - Matthew 22:17-21, NIV Bible

CHAPTER THREE

Making the Deal

In all your excitement to get your dream house underway, you may be anxious to get this part over. Spend the time to be thorough in your investigation of both your credit worthiness and the lending institution where you will get your contractor's loan. This effort can save you many dollars through the life of your mortgage.

Securing Your Lot

You may want to secure the lot of land ideal for your needs before you have found a Lending Institution to grant you a mortgage. This can be done with hand-money and a written Note stating all hand-money will be returned to you if you are unsuccessful in obtaining a mortgage within "x" days. Depending upon the market, you should be able to secure your lot for approximately 30 days before the Seller negates the purchase agreement and returns your hand-money. Give as little hand-money as necessary to secure the lot. You want to minimize your loss if something goes wrong in the process. Some land owners dictate that you must sell your land back to them instead of selling outright to someone else.

Finding a Lending Institution

Either while you are looking for the right property and / or dream home or once you have found both, you will probably need to locate a lending institution. Before accepting your first option, investigate what is available before signing or depositing any money for your loan.

Some Owner-Builder Companies have their own sponsored mortgage companies. These may or may not be the best available resources for obtaining your mortgage. Do spend the time getting their terms (interest rate, amount they will lend you, monthly pay-back, fees, and any penalties such as late fees or early pay-off fee). Compare this information with any other Lending Institution you are able to find.

We found that there were very few banks interested in Owner-Builder Self-General Contractor Mortgages. Reasons included unsure of your experience; there is a high rate of overruns; and, many do not successfully complete the job in a reasonable (~12 months) time frame. We called numerous local banks (many were found in the Real Estate Mortgage Rates Section of our local newspaper) asking if they offered construction loans for "owner / builders". After several "no's" we began asking if they knew of any lending institution that did offer construction loans for people like us. We learned of two Lending Institutions in our local area that would sponsor Owner-Builder Mortgages. One of two we immediately ruled out because they required the property to be purchased before they would consider our application. We wanted to include the property purchase in with the mortgage.

The Lending Institution we found outside of the Owner-Builder sponsored company had much better terms (lower interest, easier payment terms, and no penalties for early pay-off).

Tip

Don't be pushed into a decision to borrow money from the first lending institution you encounter. Make sure you shop around. It can be worth thousands of dollars and much better terms to you over the life of your mortgage.

Getting the Paperwork Together

Spend the time to get your paperwork in order before going for your first inquiry about a loan. We did a lot of homework up front that made getting the loan a smooth transaction. We left the Loan Company with the correct impression that we were prepared and organized, thus worthy of being responsible for the loan we were requesting.

Step 1 - Credit History

Check your Credit History. You want to be the one to know if everything is in order or if a few calls or letters can get things in order before someone who you want to borrow money from checks things out.

You should be able to get a free credit check (usually up to one per year per person) from:

> Experian Credit Information
> Box 2106
> Allen, Texas 75013
> 1-800-567-5470

Spend the time to get your credit rating checked. We found a few surprises - one being that over the years, any charge card that we had opened, never closed due to inactivity. Many years passed and we had even forgotten about particular accounts, but if they would be left on your record, it would be treated in a negative light in that you have the potential to get into this much more debt. We wrote many letters closing old accounts. This can take a few weeks to a few months so you want to do this early.

Step 2 - Construction Loan Lending Institutions

Call the Lending Institutions that handle Construction Loans. Ask them to send you their current Loan Offerings and Rate Information Data Sheet; their terms for money dispersements and repayment of the loan; and, an application for a construction loan. Having these items will allow you to compare interest rates and terms, how the money will be dispersed to you and what you will need to have ready for the loan application.

Check to see if it will be acceptable for your Sub-Contractor to be paid after all the work is complete (per Sub-Contractor segment). Some Sub-Contractors want a certain percentage of payment up front at the start of their job.

Step 3 - Your Assets and Liabilities

Get your current assets and liabilities paperwork together. Pull together your most recent bills and income receipts / records. The Bank will want to see your current assets (income, savings, investments) and liabilities (credit debt, car loans, mortgage or rent statements, other outstanding loans, child support, lease obligations, daycare expenses, taxes, and typical monthly expenses).

We went to the Lending Institution prepared with an extra photo copy of each of the following:

- most recent pay stubs

- current checking statements

- current savings statements

- most recent bill / loan statements

- Federal Tax Records for the previous two years

The Lending Agent was impressed with our organization and ease of doing business with us. We were able to secure our loan without any problem.

Tip

Spending some of your own time preparing and organizing all your paperwork before your first meeting with the Lending Agent will pay dividends in helping to secure your construction loan. You will be sending the message that you are organized with details which is just the sort of thing that someone is looking for in determining whether to lend you money or not.

Step 4 - Cost Analysis Estimate

The Lending Institution will want to see your estimate of the cost analysis and description of materials for building your house as part of their review before granting you a loan. A blank example of each of these **Cost Analysis / Budget for your Owner-Builder Home Worksheet** and **Owner-Builder Home - Description of Materials Worksheet** are at the end of this step.

Try to get at least two estimates for any work you will be sub-contracting. We found this was next to impossible for a few types of work. We began this effort during a busy period (mortgage rates were low and new housing was plentiful). In one case, after we had called the Sub-Contractor several times for and estimate to excavate the property, we finally received a bid which was almost ten times what we had estimated the cost would be. We didn't call that Sub-Contractor again.

The Sales Person who sold you the Owner / Builder Plans will have prepared a Cost Estimate of materials and labor for you. They are familiar with this work but we found that they **do tend to underestimate items**. These underestimates were largely due to unforseen mishaps.

It's best to build some cushion in your budget. We are sure you will need it. If you do happen to come in under budget, the money you haven't spent is money you haven't borrowed. You will be ahead with your construction loan - or - you can use these funds for finishing touches like draperies, wall paper and decorations.

For wisdom is protection just as money is protection. But the advantage of knowledge is that wisdom preserves the lives of its possessors - Ecclesiastes 7:12, NAS Study Bible

Tip

If your lending institution has concerns about any of your Sub-Contractors or the company you plan on purchasing your house plans from, take the time to investigate their concerns. They are in a good position to know if you should be concerned too. The lending institution may still approve your loan, but you may save yourself future headaches if you check out any concerns before you purchase the plans or hire the Sub-Contractors in question.

Cost Analysis / Budget for your Owner-Builder Home

Item	Material Cost	Sub-Contractor Cost	Total
Permits			
Inspections (Framing, Plumbing, Electrical)			
Tap In Fees			
Well, Pump, Septic System			
Land Survey			
Site Preparations			
Excavation			
Temporary Utilities Installation			
Footers			
Foundation			
French Drains			
Floor Drains (Basement and Garage)			
Posts and Beams			
Basement Floor			
Cement stairways			
Shell Framing			
Rough Plumbing			
Roof			
Windows - Exterior			
Doors - Exterior			
Trim - Exterior			
Painting - Exterior			
Fireplaces - Log Burning			
Fireplaces - Gas			
Rough Electric			
Rough Heating and Cooling			

Cost Analysis / Budget for your Owner-Builder Home

Item	Material Cost	Sub-Contractor Cost	Total
Brick Laying			
Siding			
Soffit, fascia and down spouts			
Insulation			
Interior Walls			
Ceilings			
Interior Painting			
Finish Electrical			
Finish Plumbing			
Flooring			
Finish Heating and Cooling			
Kitchen Cabinets			
Counter Tops			
Kitchen Built-in Appliances			
Bathrooms - Cabinets			
Built-In Vacuum System			
Built-In Alarm System			
Interior Trim			
Doors			
Porches			
Decks			
Sidewalks			
Landscaping			
Driveway			
Insurance - Home			
Insurance - Mine, Flood, or other			

Owner-Builder Home - Description of Materials

Item	Material
Excavation - Soil Type	
Footings - concrete mix, strength, any reinforcement	
Foundation Wall - material, any reinforcement	
Interior Foundation Wall - material	
Columns and girders - material and sizes	
Piers - material, any reinforcement	
Foundation - Waterproofing and termite protection	
Basement less space - ground cover and insulation	
Chimney and flue lining material	
Fireplace and heater flue size	
Chimney Vents - material and size	
Fireplaces - type and ash dump / clean-out	
Fireplace facing, lining, hearth and mantel	
Exterior Walls - wood frame (wood grade and species)	
Building paper or felt	
Sheathing, thickness, width	
Siding, grade, type, size, exposure, fastening	
Shingles, grade, type, size, exposure, fastening	
Stucco, thickness	
Masonry veneer	
Exterior wall sills	
Lintels	
Masonry - solid, faced or stuccoed	
Total wall thickness	
Door sills	
Window sills	
Floor framing - joists (wood, grade, and species)	

Owner-Builder Home - Description of Materials

Item	Material
Concrete slab flooring	
Floor reinforcement	
Fill under slab, thickness	
Subflooring material (grade, species, size and type)	
Finish flooring	
Partition framing - studs (grade and species)	
Partition framing size and type	
Ceiling Framing joists (wood, grade and species)	
Roof Framing rafters (wood, grade and species)	
Sheathing (wood, grade and species)	
Roofing underlay, weight or thickness, size and fastening	
Flashing material and gage (or weight)	
Gutters and down spouts, material, gage, size and shape	
Lath and plaster - walls and ceilings	
Drywall - walls and ceilings	
Joint treatment	
Interior doors - type, material, thickness	
Interior trim - type, material, thickness	
Windows - type, make, material, grade, sash thickness	
Screens - type, number, screen clothe material	
Entrances - main door material, width, thickness	
Other entrance doors - material, width, thickness	
Head flashing, weatherstripping type	
Screen doors - thickness, number, screen cloth material	
Shutters - hinged or fixed	
Exterior railings	

Owner-Builder Home - Description of Materials

Item	Material
Exterior millwork - grade and species, paint, no. of coats	
Kitchen Cabinets - material, lineal ft. of shelves and width	
Kitchen counter top - material, edging, back splash	
Bathroom Cabinets - material and width	
Medicine cabinets - make and model	
Stairs - basement (material and thickness)	
Stairs - main (material and thickness)	
Flooring - Kitchen (material, color, border, sizes, gage)	
Flooring - Bath (material, color, border, sizes, gage)	
Plumbing - Fixtures (number, location, make, size, color)	
Heating - type, make, model	
Electrical wiring - Service (overhead or underground)	
Electrical wiring - special outlets (items and location)	
Lighting fixtures - number of typical	
Lighting fixtures - number non-typical	
Insulation - location, thickness, material, vapor barrier	
Porches	
Terraces	
Garages	
Walks (width, material, thickness)	
Driveways (width, material, thickness)	
Landscaping - topsoil	
Landscaping - plantings	

Step 5 - Waiting on the Loan

While you are waiting for the Lending Institution to approve your loan, there are several things you can start:

- Contact Sub-Contractors

You can compare Sub-Contractors bids, making sure you get full cost estimates of materials to be supplied by them or you, lead times and completion times. A blank **Sub-Contractor Bid Request Sheet** can be found a few pages later in this chapter which you can use as a guide in sending out for your Bids. We have also prepared a number of "completed" Sub-Contractor Bid Request Sheet examples, located in Appendix 1, to use for ideas of what you may want to included in getting Bids. Remember to consider local Sub-Contractors. They are in our neighborhood so it will be easier for you to find out their business work history. Don't forget to check your local Better Business Bureau for any Sub-Contractor's reputation. Also, local Sub-Contractors may be able to work on your job even if the weather was initially bad (if their job depends upon the weather). Since you are close, time is not wasted getting to your site.

Also remember that since most Sub-Contractors' business is generated by word of mouth, it's a good idea to start mentioning your home building project to friends, family members and co-workers to get ideas and recommendations of Sub-Contractors' work.

Here's an extensive list of Sub-Contractors you may need to send Bid Request Sheets to:

Well System (if required)

Perk Test

Surveyors

List of Sub-Contractors you may need to send Bid Request Sheets to - continued...

Clearing / Hauling

Excavation / Removal

Septic System

Dumpster Service

Portable Toilets

Footer Installation

Block Installation

Framing

Brick Installation

Concrete Supply

Block Supply

Brick Supply

Concrete Finishing - Internal

Beams / Lintells Supply

Roofing

Siding

Exterminator

Electric

List of Sub-Contractors you may need to send Bid Request Sheets to - continued...

Fireplace

Heating

Plumbing

Gutters

Garage Doors

Insulating

Drywall

Painters

Trim Carpenters

Carpeting

Glass Showers

Appliances

Brick Wash

Asphalt

Concrete Finishing - External

Landscaping

Painting Supplies

God has made everything beautiful for its own time. He has planted eternity in the human heart, but even so, people cannot see the whole scope of God's work from beginning to end. So I concluded that there is nothing better for people than to be happy and to enjoy themselves as long as they can. And people should eat and drink and enjoy the fruits of their labor, for these are gifts from God. Ecclesiastes 3:11-13 - NLT Study Bible

Sub-Contractor Bid Request

To (name of company):
Name of Contact Person:
Address:

Phone Number: Fax Number:
e-mail address:

From:
      (Owner and General Contractor)
Address:

Phone Number: Fax Number:
e-mail address:

Subject: Bid Request for:

Work Required by Sub-Contractor:

The Sub-Contractor will submit an itemized bid for the following work:

Sub-Contractor's Certificate of Insurance (must attach a copy with this Bid)

Materials supplied by Owner / General Contractor:

Materials supplied by Sub-Contractor:

What needs to be completed before the Sub-Contractor arrives:

Lead Time: Completion Time:

Start Date: Expected Finish Date:

Special Items:

      Please submit your bid in writing within 2 weeks of receiving
      this request of bid to the above address or fax.

Note: We have given more detail to the importance of Sub-Contractors in the next chapter.

- ## Contact the Local Municipality

Get the application for the Building Permit. Also, find out if any Building Inspections are required and what lead time is necessary for the Inspector. Find out what your fees will be for tapping into the water and sewer (or equivalent).

- ## Contact the Utility Companies

A Notice of Intent to Construct Form must be completed before receiving power. Obtain the appropriate form(s) from your electric company. Remember that YOU are the Builder so your name goes on any forms. Contact the Gas Company too for an application.

- ## Contact an Insurance Company for Home-Owner's Insurance

You will need to supply proof that the Homeowner's Insurance is adequate for the construction project, and that it's paid for (typically paid for one year in advance).

Make sure you check your Insurance Deductible as to the pros and cons of having a higher or a lower deductible. With this being our first years, we kept our deductible low. Anything can happen when your are building a house. The up front cost of having the low deductible was worth it for us in having the comfort of getting reimbursed if something gets **stolen or damaged** (unfortunately this can and does happen).

- Special Insurance Needs

Also check your building area for special insurance needs, such as flood, earthquake or mine subsidence insurances. Our property is located in an area known for coal mining. We opted for additional Mine Subsidence Insurance.

Tips

Supplies can and do arrive at any time. Plan for their arrival by having areas designated for each. Place a sign or post exactly where you want the supplies to be placed / dumped in case the delivery is made when no one is around.

Protect the wood and windows that are going to be stored outside until installation with something like dark plastic (protects from weather and view).
Renting a trash bin early (framing stage) will reduce the mess you will have to pick up later. This will allow the Sub-Contractors to keep their messes to a minimum. Trash bins are expensive so try to anticipate when they will be required. The drywall waste will be the largest mess.

If your local township permits open burning, you may elect to save some money by regularly burning the excess wood and trash.

For we know that if the earthly tent which is our house is torn down, we have a building from God, a house not made with hands, eternal in the heavens - 2Corinthians 5:1, NAS Study Bible

CHAPTER FOUR

Sub-Contractors can be Your Best Decision

Bid Requests should contain a lot of detail to enable the Sub-Contractor to make an accurate bid. The nature of the work should be described to include what you want done, where it is to be done, what materials will be supplied by you and what you expect to be included in the bid. Include a diagram or drawing if appropriate. All materials to be used should be specified in detail. If appliances are to be included, list the brand name, type, model, number, color and size for any item. And do not forget to obtain a copy of the Sub-Contractor's Certificate of Insurance. It should include the Insurance Provider's Company Name; Insurance Type; Policy Number; and, Expiration Date. Knowing your Sub-Contractor has the necessary insurance will protect you should an accident happen.

The company where you purchase the plans for your Owner / Builder house should supply you with a list of Sub-Contractors that they recommend. These Sub-Contractors are folks who should have worked on these homes and would be familiar with the types of plans that the company sells. This is a good starting point for where to send bid requests for different job assignments.

Cost, reputation and availability are factors that you have to weigh carefully when evaluating a bid. You have to consider too if only one Sub-Contractor has placed a bid, do you want to accept or take time to try to get others. We had a few situations where only one Sub-Contractor submitted a bid to us.

Tip

We found that we could purchase some items cheaper than some of our Sub-Contractors. Don't forget to tell your local Home Center that you are acting as your own General Contractor. You could get a 10% Contractor's Discount that would not typically be available to the general public.

The Contract

Where possible, get it in writing. The agreement between you and the Sub-Contractor you hire can be as simple as a handshake or as formal as a contract. Communication between you and the Sub-Contractor is the most important ingredient between any deal. When you get it in writing there is less room for a misunderstanding. However, if your contract is too specific, the Sub-Contractor may resend the bid and refuse to do the job. Also remember that the best Sub-Contractors seem to be the hardest to get during the peak building season. If you do find a Sub-Contractor that is easily available during peak season, you may be a little skeptical of the quality of work that you may see.

Although we were unsuccessful in having a start and ending date agreement with our Sub-Contractors, this would be good to get. Remember that even if you agree to firm dates, uncontrollable circumstances can still occur to alter the agreement. Weather was a culprit. When it rained or snowed it usually pushed all the Sub-Contractors we had scheduled back. Do not lock yourself into a contract that you may have to break and could cost you money or time to get it straightened out.

Make sure you are clear with what is and is not included in the Bid from the Sub-Contractor. We were getting bids for having the basement and garage cemented. We were pleasantly surprised to see that one Sub-Contractor who had an excellent reputation for his work was well within the other bids. We gave him the job. The work was excellent. Our surprise came when he handed us his bill and the bills for all the cement he used. This unexpected expense put us over our budget for these two items (basement and garage cementing). From that point on we knew to always ask what was and was not included.

Another thing that happened to us of not having something included in writing occurred with our gutter installation. We assumed that the gutters would be cleaned at the end of the job. We were wrong. We were able to see into the gutters for our garage and noticed how filthy these were. The job was already paid and the Sub-Contractor was on to another job elsewhere. Rather than wait for him to come back as his schedule would permit, we went ahead and cleaned them ourselves. We did not want water overflowing from all the dirt, leaves and debris. Another lesson learned!

Paying the Sub-Contractor

A schedule of payments should be agreed upon with the Sub-Contractor. Insist that the payments do not get ahead of the work actually completed. Remember, once the Sub-Contractor has your money you have very little leverage.

When you receive a bill or statement from your Sub-Contractor, check for:

- an itemized list of what you are paying for. See that it matches what you see is in place.

- compare this list with your original quote. Ask questions if there are any discrepancies.

- make sure you are not paying twice for things or that you are not paying for items you supplied.

- if possible, make daily trips to the house to check on the completion of the Sub-Contractor's work. Daily trips will allow you to not only check on the work status, but also allow you to keep the work area clean. If you make an effort to keep the area clean, the Sub-Contractors will sometimes be neater knowing that is how you want the area kept.

Learn from your Sub-Contractors

Stay open to suggestions from your Sub-Contractors that are good and you trust. Ask questions to find out as much as you can about the details of the work they are doing. The more you learn from your Sub-Contractors, the better decisions you can make on how you build your house. We made many changes and alterations for the better along the way because of sound information we received in talking with our Sub-Contractors.

Keep your records organized

We had to provide copies of receipts for the work completed or items purchased to build the house for proper reimbursement. To keep everything organized we used a system of having all receipts that had not been sent to our Lending Institution placed into a shoe box. So if we had just gone shopping for house stuff or paid a Sub-Contractor we would take the receipt and immediately put it into the "receipt" box. When it was time to send for a dispersement check, copies of whatever receipts were in the box were made and sent as our proof of what had been done or purchased. The original receipts were then transferred to a folder that was marked "sent to bank". As crude as this may sound, it was easy for us to keep track of what had and had not been submitted for reimbursement from the bank.

We also found this system was handy when needing to return items that were not necessary, wrong (size, color, model, make), or damaged. We knew the receipt was in one of two places. And we kept every receipt! Whatever system you come up with, make sure you implement something to keep your receipts organized. The key to a smooth process of building your house is to keep your records organized.

Tips

Get a mailbox (the plastic type can be found to be rather inexpensive) and place it on the property specifically for your Sub-Contractors to use to communicate with you. They can leave messages or receipts for your attention. The mailbox flag can be placed in the up position to notify that you have mail from a Sub-Contractor.

And some sad news...

Keep in mind that because you are only building one house, you have very little clout with Sub-Contractors. You are a one-time deal. When a regular General Contractor calls the Sub-Contractor you have hired, expect that they will drop everything for the regular General Contractor. These General Contractors are probably your Sub-Contractor's regular paycheck. As jobs come up, that Sub-Contractor will not want to miss out on future work if they turned down something while working on your house. We found many times the Sub-Contractor would drop our job for days or weeks to keep someone else happy. It was frustrating to see our Sub-Contractor busy on another job while our job was still unfinished. Getting something in writing (if you are successful at this) may help keep your job moving along at a planned pace. We were not always fortunate.

When you build a new house, make a parapet around your roof so that you may not bring guilt of bloodshed on your house if someone falls from the roof. - Deuteronomy 22:8, NIV Bible

CHAPTER FIVE

Let's begin that Dream House

When you are in discussions about the blueprints for your particular model, request that they contain your model elevation and options only. Our blueprints contained four different elevations (optional variations of the basic model). This caused us so many problems. The Surveyor erred by four inches on our house layout due to not following the proper elevation throughout the blueprints. This resulted in our lot not being properly excavated the first time (which equals more expense when you have to call them back several times). The final frustration with all of this came when we were told the framers could not set our house on the footer and block - that it was eight inches off square. We finally got this all straightened out, but if we had better blueprints from the start, none of these frustrations would have occurred.

Most Owner-Builder Programs offer packages which include all the framing materials. We opted to purchase the framing package which has preformed walls. Since they are preformed, the foundation must be laid out perfectly to accept the walls.

If you want a mirror image of your house, be especially careful to check all the details on the drawings. Do not rely on the Building Company to have everything correct. It may be worth the money to have a professional house draftsperson review the drawings supplied by the Building Company. This is the best time to find errors, not when the house is under construction.

One of the most frustrating considerations of home building is that the overall progress depends on the sequential completion of tasks (many of which others perform and are not completely within your control). And if the materials aren't at the job site when the workers are, no one gets any work done.

Tip

If you do receive blueprints with more than one model elevation and options, spend time to highlight / clearly mark your elevation and options. Do this for every set of blueprints you will be using. Clearly explain to everyone using your blueprints which option is yours.

Survey the Land

Contact a local Surveying Company to survey your property. This must be done before you do anything else to the property. They will be out for a few survey items throughout the building process. These include:

- Survey your lot

- Set property corners

- Prepare plot plan for Building Permit

- Stakeout the house for excavator on offsets along side property lines

- Final building location (as-built)

- Prepare a certified plot plan to be used for occupancy, the mortgage co., etc...

Get your Permits in Order

If you haven't already contacted your local municipality for the application for the Building Permit, do this now. You MUST have a Building Permit displayed on your property before you can begin.

Submit your Notice of Intent to Construct Form to receive electricity from the local Power Company. This will be important for any Sub-Contractors who have power tools or for any temporary security lighting you may put in place.

Preparing / Clearing the Property

Once you have Closed on your Construction Loan, your Lending Institution will notify you when you can begin clearing the property. They typically want to check the property right after closing and get a few photos for their files on your property. Make sure you find out how long you must wait before starting to clear the land. They make take two to seven days to complete their inspections.

Plan to clear as much of your property as you need to clear before you begin building. It will be much more difficult and could result in damage to your house if you wait to do some of the tree clearing after construction.

When we set out to clear our property the first thing we did was purchase a good chainsaw with an extra blade. Having the right equipment means the difference between a safe, controlled working environment and an accident waiting to happen.

The Surveyors will stake-out the house position on your land. Use this staked area to figure what trees need to be removed. Leave at least a 3 foot stump when cutting down your trees. This will allow the Excavator some leverage in pulling out the stump.

Dennis using a come-along to secure a tree for cutting.
One thing to watch out for is the poison ivy!

Tip

In the process of clearing your land, select one or more trees on your lot that can be used for lumber. Arrange to have these trees cut for use in your house. This will be a nice conversation piece and a remembrance of the land you cleared.

Excavating

Excavators typically charge by the hour. Their rates vary depending on the size of the equipment used and what you want done. Review with the Excavator the estimate of work to be completed. State in writing if you want certain trees, stumps and excess dirt removed. Make sure you indicate or mark the trees you want saved.

When excavating the land, save that valuable topsoil. Have the excavator put it in a pile on your land. It will save you from purchasing topsoil later for your landscaping. Before the excavator sets the elevation of the home, make sure the sewer depth and utility lines are known.

Remember to have the land removal include clearance for all necessary items such as the French Drain and any special piping or drains that may lay outside of the footer boundary. Now is the time to get all the earth removed for all your construction needs.

Some problems you may run into while Excavating include:

- Rock surface before reaching your desired depth

- In climate weather, delaying the work

- Springs

- Excessive digging which requires back fill

At this point you might be able to have a water line run and connected for your water supply. You will need water for various things early in the construction, including water for the footer and block work. Other options for water can include renting a temporary plastic water tank or, if you have neighbors close by, asking them if you may use their water during some of the construction phase.

This is a good time to have the Excavator spread limestone or gravel for a dry lay down area after the excavation is complete for your basement. You will have to call your local gravel supplier and coordinate the delivery of your limestone to the completion of your excavation. The Sub-Contractors appreciate a dry area to park, load and unload their tools.

Make sure to check the excavation for level. The Excavator will have a transit. You can have the land verified that the excavation is within 1-2" of being flat.

Access to the property containing several truckloads of 2B limestone.

Make sure you have prepared a section of property for the framing materials

The Framing Materials will arrive typically in one to three bundles, depending upon the size of your house. The preparation to be done before this arrival includes:

1. Have a clear spot on the property near the house foundation that is large enough to fit the number of bundles you are expecting. This land should be clear of trees, trash, and dirt. It should be easily accessible from the road. Our lumber arrived on flatbeds the size of a tractor trailer.

Two of the three bundles of lumber that we received.
It is resting on the limestone that we had prepared for its delivery.

We spread limestone on the ground where the bundles were to be placed. We did not want the wood to get muddy or warped. The limestone helps to keep this area free from mud and puddles.

2.   The path to the materials storage area would be best if it were level to the framing material drop off point. Carefully consider any graded path because the trailers hauling the lumber and supplies may not make it to your cleared area. This path would be a good area to have stones spread for ease of entry and exit from the cleared storage area.

You may end up having your framing materials a distance away from your foundation if the delivers cannot safely and easily drive onto your land.

3.   You may also want to place a sign indicating the area where you want the supplies to be placed in case no one is available at the time of delivery. The sign can be as simple as spray painting the location / ground of where to drop the items or spray painting an old piece of cardboard or wood and posting it in the general area.

4.   The Drivers will want to drop the load and move on. Make sure to visually inspect as best that you can for any gross damage. A Representative should also be there when the bundles of framing material are disassembled for an inventory of all the pieces.

Have a secure place for windows, doors, door fixtures, and shingles ("walk-off" items)

The packaged house we purchased arrived including the doors, windows, door fixtures and shingles - items which could easily "walk-off". We were not expecting this to arrive at the same time as the framing supplies and was not prepared to secure these items. We did place our windows aside and covered them with dark plastic that blended in with the scenery. Fortunately, no one bothered the area and the windows were installed without any losses. The door fixtures were easy enough to load on our truck and take home.

We shared this news with other folks. They were able to be better prepared in securing these types of items. One family rented a small trailer which locked. They kept this trailer on their property until the items were installed. Another family had relatives in the area and were able to use their garage and basement areas while they were building.

Here are our windows covered with black plastic to blend in with the surroundings and be less obvious from the roadway.

Have designated places for other supplies

Once the house gets underway other materials will also begin to arrive such as block, brick, sand, dirt, rock, stone, etc... To ensure that these items are placed where you plan for them to be placed, in addition to notifying the appropriate suppliers / deliverers, again, make signs with the item listed and place the sign in the area you wish the item to be unloaded. This could save headaches later of having material dropped in a place that would be better suited for an item that was to be used first.

Ask the Sub-Contractors where they would like these materials placed. Our Bricklayers wanted cement bags and sand as well as the brick located in several areas around the house construction.

For ground that drinks the rain which often falls upon it and brings forth vegetation useful to those for whose sake it is also tilled, receives a blessing from God - Hebrews 6:7, NAS Study Bible

---

Tip

Do not assume that others will look for the most logical place to deliver any of your supplies. They will typically look for the easiest drop-off point if you do not give them some direction on where to place specific items.

CHAPTER SIX

Beyond Excavation

Once you have the site cleared and excavated, you're ready to start building that dream home, beginning with the critical footer and foundation. Your house will sit on these so it's important to you to have them be installed correctly.

You will learn that each step is its own separate process, from getting the Sub-Contractor to arrive on time, to checking the work and determining if it meets the prints and your standards.

Footer

The lowest part of the house is the footer. This is were the whole house rests, so it's extremely important to have this sit on solid earth. The frost line must be determined for your location. The frost line is an imaginary line below the surface where the ground freezes. Below this line is frost free. If a footer is not placed below the frost line, the typical expansion and contraction of the earth might cause the footer to move.

Separate piers should be poured wherever steel support columns will be located. These piers will be detailed in your blue print house plans, showing the placement and size for each pier.

These support columns will secure the steel beams used to hold the inner floor joists of your house. Also make sure you include clearance space for placement of your French Drain next to the footer.

When we talked to the Footer Sub-Contractor we asked him to "beef up" the footer for more strength since the entire house will be resting on this. Our plans called for a 20" width by 8" thick footer. We asked the Sub-Contractor to make the footer 24" wide and 11 ½" thick, reinforced with steel rods. After the footer was poured, we had 2B limestone trucked in to put a 4" to 6" base inside the border of the footer. You will need to have stone later for the basement floor but doing this step now helps to keep the excavation dry and mud free while the Block Layer Sub-Contractors build the foundation.

Remember to put a through pipe in the footer to allow access for the sewage line later.

Wooden forms for the footer.
This is the first step in preparing to pour the footer.

The footer has been poured and
the French Drain has been partially installed.

French Drain

A French Drain is fairly simple and inexpensive to add. Make sure if you do not know how to install one, that you get someone who is knowledgeable in installing these. Install the French Drain just after the footer forms have been removed. We used a 4" plast coiled drain pippe. After you place it down, cover it with 2 to 3 inches of 2B limestone to keep it in place and to protect it while the foundation is being built.

A French Drain will aid in keeping water from entering into the low portions of your house and foundation. This is especially important if you experience freezing weather. If the water cannot escape, it may freeze and cause damage to your foundation.

We decided to fill in and around the foundation with 2B limestone. We stopped 1 ½ feet from the top of the grade. This will allow water to quickly find the French Drain and be channeled away from the house.

Foundation

The foundation is what sits directly above the concrete footer and below the sill plates which the floor joists will sit on. This is typically referred to as the basement walls. Concrete block has been the most typical material for the foundation walls but more homes are being constructed with cement poured walls. Both offer advantages and disadvantages. The cement wall overall is stronger but once a crack forms, it will usually travel a greater distance than a wall made with concrete block. Concrete block is cheaper and easier to install.

The block was delivered inside the excavated area where
it sat on a limestone bed

It was suggested to us to have the same person who installs the footer also install the foundation. This will ensure that the foundation will properly fit to the footer. Since the same person will be installing the wall (block or cement), they have a vested interest in laying the footer correct to the prints. We did do this and agree that it was worth the effort to coordinate, especially since we had some problems initially with getting the lot properly surveyed with the house prints. The Sub-Contractor understood the difficulties we were having and helped us through this situation.

As we did with the footer, we also had the Block Sub-Contractor use 12" wide block rather than the standard 10" block. The price difference was only a few hundred dollars to use the larger block. We had the block delivered inside the excavated area where it sat on the limestone bed (see photo).

Another extra we asked for was the use of tie down bolts. These bolts are cemented into the last top course of blocks and will be used to hold down the sill plates which will be placed on top of the foundation wall. Because we wanted the bolts placed in the foundation, the termite block (which is a block that the cavities are ½ filled) were not used in the area where the bolts were placed. In these areas we used a normal block and filled the cavities with cement and set the bolts.

Your plans should give you the diagonal measurements. They will be circled alpha characters at various corners. Take a 100' tape measure and measure each marked point to determine if your foundation will match your pre-build walls.

Once the foundation is complete consider having an Exterminator come out and spray the walls for termite protection. All this ground movement could cause activity for the local insects, including termites. You will want to address this issue before it turns into a problem.

Make sure the Footer and Foundation match your Blue Prints

As we began to mention in the last section, something we did not expect to happen was that the foundation did not match the blue prints. This quickly caused panic in our hearts when we heard the framers tell us that the house was not "square" and that the back sun room was too short. This was a surprise to us because we even went and measured the model home that was identical to the house we were building. The error of the sun room being slightly short occurred when the Surveyor got confused with the blue prints. The marks he made were really matching two different elevations on the blue prints that we received from the Owner / Builder Company.

We are still not completely sure how the house became off-square. Fortunately, our brick layers were familiar with this problem and easily corrected it by bumping the bricks out and providing extra support in the un-square locations.

We strongly recommend you either ask for prints containing only your house design, or if that is not possible, highlight the correct elevation on the blue prints. Clearly state to any Sub-Contractor which elevation is your house if there are more than one style included in the prints.

Prepare the Walls for Waterproofing

We have seen various methods for waterproofing your walls. You may want to check with your local Home and Garden Store for recommendations on what works best. We used a black tar-like substance painted on the block followed by placing thick black plastic then sheets of styrene foam for insulation and weather protection. We have included two photos showing the waterproofing.

A house is built by wisdom and becomes strong through good sense - Proverbs 24:3, NLT Study Bible

The Waterproofing Process of adding the
tar-like substance to the cement block.

Here is the underground wall covered with the tar-like substance
(before the black plastic and foam sheets).

Beams

Now that the foundation wall is up we will need to set the beams and support columns. For this job you will need to contract a Crane Operator and crane. In your foundation will be pockets left open where the beams will rest. When you order your beams make sure to also order some metal shims that will be required to shim the beams into the beam pockets to level.

Here is Dennis unloading the beams from the deliverer.

Your plans should show the location of the support columns for each beam. The piers that were poured when the footer was poured will give a solid footing for the support columns. The framing package that we purchased included the support columns. Secure the beams and columns with 2 x 4 lumber (see photo below showing the beams secured with lumber).

The Beam is inserted into the beam pockets and
secured with 2x4 lumber and metal shims.

Framing the House

Most Framing Sub-Contractors will ask to have at least on side of the foundation to be back-filled to allow access to and from the house for framing. You will need clean fill material and a means to put it around the foundation.

The company that sold us our house package and framing package also offered the services of their Sub-Contractor Framers. It was easy to see it would be a wise investment for us to go with a crew that was already familiar in building the model of our home. The price for framing from that company was very competitive.

Remember that during the framing phase is NOT the time to make major structural changes. Significant changes need to be done while you are "Making the Deal" with the Owner / Builder Company. We were able to make few minor changes (like adding a doorway and raising the garage door height from 7 feet to 8 feet).

A plastic foam material was rolled out on top of the block foundation to reduce water from the block to wick up into the wood sill. Then a sill plate was placed on top of the foam. The wood sill place which contacts the foam should be treated lumber to resist bug infiltration and damage from water. On top of the treated sill a standard wooden sill. These were both bolted down using the bolts cemented in the top of the foundation. The Framing Sub-Contractors made the final adjustments to the beams and support columns. The beam should be adjusted to be at the same elevation as the treated sill plate.

The single and double wooden sill plates on the foundation wall.

A 2" x 6" piece of lumber was placed on top of the beams which brings the untreated sill plate and the 2" x 6" on top of the beams to the same elevation.

The Framing Sub-Contractors laid the floor joist on top of the sill plate and the 2" x 6" on top of the beams (see photo below).

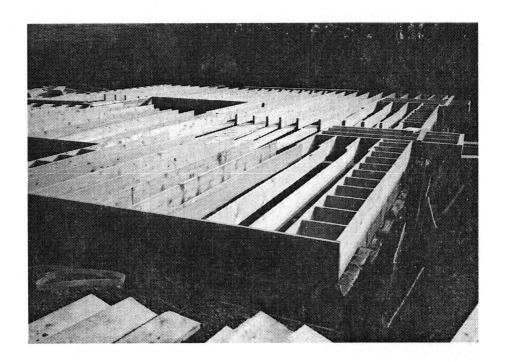

Floor joists on top of the foundation.

The decking material was then placed on top of the floor joist. Each sheet of decking was glued and nailed in place. The glue will help reduce the squeaking floor problems that occur with many houses.

Tip

When the Framers started we rented a portable toilet facility. Our Framers thanked us for this "pleasant" convenience. We were told by some experienced home builders to budget for such a convenience, otherwise you may find human waste in undesired locations!

Floor decking in place.

Most framing packages will have pre-assembled walls. Because they are pre-assembled the foundation needs to be perfect. Raising the walls is one of the quickest parts of the whole building process. You will be amazed from one day to the next how much progress seems to be gained.

Walls going up during the Framing Stage.

During the framing stage you will want to keep an eye on the construction and your plans. Especially if you made any changes to the original plans. We had changed our Master Bathroom layout. Even after going through these changes with the Framing Supervisor it was not originally laid out the way we specified. We had them rip out what they had done and re-do it to our desired, agreed upon specifications. This will cause delays so it is best to keep on top of the project to ensure you get what you expect to receive.

Tip

Purchase some large plastic tarps to cover up your wood which will be exposes to the elements. We were told that our framing material would arrive wrapped in plastic. It was not wrapped! These bundles are as large as a tractor trailer so a few tarps to shield your material is a must. Tarps may also be necessary to cover the roof to keep water (rain or snow) from entering before the felt paper or shingles are installed.

Final Walk-through with the Framers

This is the last time that things can conveniently be brought up as far as changes with the house before the drywall is in place. You will probably have been monitoring the framing process all along while they were working. If you have not been doing this, than this final walk-through is a critical step.

Things you should review during the final walk-around with the Framers:

Are the correct number of windows in the correct locations?

Are the correct number of doors in the correct locations?

Were there any special changes requested and were they done as requested?

Are all the rooms done to the drawings that you purchased and agreed upon?

Were the correct materials used?

Check to make sure the walls and ceilings are true enough to allow the Dry Wall Sub-Contractor to uniformly cover them.

Are the windows plumb and level in the openings? Are they secure?

Is there enough of an opening in critical areas where you will be bringing furniture and other large pieces into the house? We found that a 36-inch opening is good enough for most large items.

Assume nothing. Check to see if the steps are nailed down and that the doors are nailed in.

Measure the rooms for square (measure between corners).

This is a good time to add extra wood between the wall studs for more support for things such as kitchen cabinets, mirrors, railings, toilet paper holders and closet racks.

Secure your House during Construction

Early in the process we had one experience where a few spools of wire disappeared. This was the best warning for us to secure our materials and tools in a locked location.

Basement: Before the house was secure with doors and walls, we secured the basement with wood walls and a locked door. This took a little effort to set up, but nothing disappeared after we kept everything secure. It also helped that we made daily trips to the site to ensure everything was safe and intact.

House: The next item to enclose was the house. The doors were in place a few weeks before our garage doors were installed. Since we knew were going to change the house door locks, we made a few extra keys to pass out to trusted workers. When an unfamiliar Contractor would need access into the house, we would stop by the property early in the morning and unlock one of the entrance doors. We would then follow-up each night ensuring the house was locked up and secure. All along we kept the basement secured and locked.

Garage: Once we had the garage doors installed, we were able to permit access into the house by limiting who received the garage door code. Once that particular job was done, we changed the programmable security code. And again, the basement remained secured and locked while we were not present. We kept our supplies and tools in the basement whenever we were not at the house.

It's worth taking the time to plan your security system during the construction phase. You don't want to have needless delays while waiting on a re-order or an insurance settlement due to thievery or needless destruction.

Lay out Sewage Lines in the Basement

The best time to think about installing a bathroom in the basement is before you have poured the cement. It is a small expense to have sewage lines put in place before the basement floor goes down. You may decide not to put a bathroom in your basement, but it could be a good selling point to at least have it plumbed for a basement bathroom.

We also installed plumbing for laundry facilities and several wash basins in the basement. Realizing how easy it is to do this now rather than later prompted us to think of several items that we "may" want in the future.

Concrete Pouring the Basement / Garage

This was an area where we did not feel we had the muscles or the know-how to pour our basement or garage. Our garage was attached so the cement was poured at a different time than the basement.

Remember that this is your last chance to make any changes to the sewage lines, laundry sinks, drain lines (for laundry, furnace and general areas). If you are in an area where water is going to be a problem, you may want to have an internal french drain and sump pump installed before the basement cement floor is poured.

And don't forget to get the Labor and Materials in your Bid unlike us with our incident regarding the Sub-Contractor who had an excellent reputation for cement laying submitting a bid which was extremely reasonable. What we neglected to understand was that his bid did not include cement (the other bids did include cement). This was a lesson learned. We have to say though, the cement work was excellent!

Plumbing

This step starts in the early phases of home building. An interesting occurrence happened with us when we found out most plumbers will not do the job if you do not let them purchase all the fixtures (toilets, showers, tubs, sinks, faucets). We found out that this is where they can make a lot of money by marking up the supplies. We were finally successful in getting a registered plumber from our local church.

We had purchased all the fixtures in advance when we were able to get a discount on our first purchase at a local home and garden center. The Home and Garden Center we went to ended up being cheaper than the prices our plumber would be able to secure. He was shocked (and we were pleasantly surprised!). So don't be afraid to shop around (for fixtures and plumbers).

If you are going to hire a Plumber to do any major work, especially work that involves the public sewer or water main, make sure the Plumber or Contractor is licensed, insured, and bondable. Most municipalities require that Contractors post a bond whenever they work on the public sewer or water main. You will probably get this person involved before you have the house completely framed. Rough plumbing can begin once you have the footer and foundation set.

Check to see that you have adequate shut off valves to water lines that you could typically need to have repaired, or that you would want to limit the water flow.

Examples of lines that you may want to consider adding shut off valves to are:

- all sinks (kitchen, bathrooms, powder room, work sinks)

- the dishwasher

- all bathtubs and showers

- the laundry facility

- all outside water lines (to hose outlets, water sprinklers)

We opted to have a 1-inch water line brought into our house. This allowed us to have a 1-inch main line with a 1 inch manifold. Off of the 1 inch manifold we ran the 3/4" lines up to the various areas in the house. We can have two showers on, a dishwasher running, and even have a toilet flushed without noticing any change in the water flow to each of these items.

Rough Heating

We sub-contracted this work out to our Plumber. He was in the business of Plumbing and Heating / Air-conditioning. We feel he gave us a good deal by getting both jobs.

Have your Sub-Contractor review the prints for the heating and air conditioning system. Our Sub-Contractor pointed out that the plans called for a system that we were not installing. If we would have continued with the original plans, our home may have been drafty and short on air in-take openings.

The Heating / Cooling System that was first specified was a system with two large in-takes located in the open family room and sun room with individual exhausts in all rooms. The problem with this set up is that if you close any of the doors to a room you would not have any circulation of air (heat or air-conditioning) to that room.

We added air vents and in-take openings in our basement. Although we did not plan to finish the basement at this time, we wanted it to be comfortable for whatever we planned to install (currently it is a work shop and laundry area).

Our heating choice was to use two electric furnaces and two electrical heat pumps to supplement the heating and air conditioning needs. We also thought it would be a great idea to have a climate / timer controller for the house. We found out that the current temperature / timer controllers are not suitable for heat pumps. We later noticed a message stating "Not for use with Heat Pumps" on the temperature / timer controllers at the store. The reason that it is difficult to use a climate / timer control with a heat pump is that a heat pump will not be able to increase the temperature by more than a few degrees in a short time. So the main furnace will start the coil heat which is much more expensive, in effort to provide the extra heat to overcome the difference in demand temperature and ambient temperature. By the way, there was gas available in our area but we decided to go with whole house electric.

Rough Electrical

Building codes virtually everywhere require that only licensed electricians do electrical work, but many of the codes make an exception that allows Homeowners to do electrical work on their own homes. This work must still be reviewed by an Electrical Inspector.

Since wiring is not too expensive, you may want to install wiring for items that you may not necessarily put in place before moving in. We had wiring installed for ceiling fans for every bedroom. Since there are only two of us, we did not immediately go and purchase fans for every room - but the wiring is in place. We also installed wiring for four phone lines throughout the house. One phone line may be plenty for you now, but you never know when those other lines could be handy or a necessity (such as the internet - Lan).

We also took advantage of the low expense for wiring and had cable outlets installed in every room (two in some rooms). We were generous with outlets, well exceeding the minimum building requirements. Now that we are moved in, these items are a pleasant convenience that you may not want to do without. Think of other systems such as security, intercoms, or whole house stereo systems that could be connected. Our next house will have low voltage switching and programmable light switches in the plans.

When the Electrical Contractor is finished with the rough-in wiring, don't allow any other workers to cover the walls until you have had a successful electrical and plumbing inspection.

You will find that building codes really vary depending upon your state, county and local township. Contact your local Electrical Inspector and ask for the requirements for your area.

Brick Laying

We had agreed from the start that we wanted to invest in having our whole house bricked. We love how brick looks and decided it was an area where we would add to the budget as necessary. To our big surprise, when organizing this ourselves by purchasing the brick separately and hiring Brick Layers, this was a small expense. We had our whole house bricked with jumbo bricks for less than it would have cost to have the front of our house bricked by the house manufacturer who also sells completed houses with brick as an upgrade option.

Consider ordering extra brick than what you estimate that you will need. We thought we ordered several hundred extra brick but found that after the Brick Layers were done, we did not have many whole bricks left over.

Knowing we had ordered extra brick, we took advantage of a few upgrades that the Brick Layer suggested including decorative brick coins on all of the corners on the front of the house and decorative borders around the windows and doors. We did not hang shutters after we saw how nice the brick borders looked. This will result in lower maintenance with not having to paint, clean, or replace shutters.

We also had a brick pattern placed above our front entrance door rather than a wood piece scheduled to be inserted between the door and large overhead window. Again, there would be less maintenance in not having to treat or paint the wood insert. And the brick blends in with the house so well.

We spent a lot of time going through different neighborhoods looking at different brick styles, colors and sizes. We contacted a local brick supplier in our area and visited their showroom. Ask questions when you contact the brick suppliers. The more you know the less surprised you may be later.

We selected a number of colors and styles of brick that were "possibilities". The Salesperson then took us for a ride showing us homes that had these bricks used on them. We were amazed at how terrible some of the bricks that we selected looked on a house. Since we wanted a lot of decorative brick work, we noticed that light color brick seems to remove the effect of the expensive brick work. We finally found a brick color and style. We also decided to use a jumbo sized brick. The brick is larger therefore it takes fewer bricks to cover your house. The Brick Layers do add a slightly higher cost for laying the jumbo brick however there was a savings of purchasing less jumbo brick from the brick supplier. The overall cost was within $500 of using either style of brick (jumbo or standard). Standard size brick will allow you to do special patterns like herring bone where jumbo bricks are too large to be used for anything more than the traditional brick work.

Tip

Take the time to attach inexpensive plastic to anything which you do not want mortar spilled on - including windows, doors, vents and fixtures. You may also want to place wood over the windows. Bricklayers some times drop bricks that might hit the window ledge and cause large dents or break windows.

Siding

If you do decide to have siding, one of the biggest challenges will be in selecting the siding material and color. Siding is available in aluminum, vinyl and plastics. Vinyl offers a quieter wear (less noise during wind, rain and hail storms) and is easy to cut and install.

Dark colored siding will expand more than the light color siding. This may cause buckling on an area that may be wrapped in aluminum. We found this to be a problem with the few areas that were sided in dark brown.

Note: If you are having both brick and siding, brick first then have the siding installed later.

Roof Shingled

The package we purchased came with shingles. We upgraded the ridge roof vents. We also had the Roofing Sub-Contractor use a special ice guard. This is a roll approximately 3 feet wide and 50 feet long. It has an adhesive backing and helps to prevent damage to the eves when ice forms in the gutters and migrates up the roof.

One of the best sources for a Roofing Sub-Contractor are recommendations from friends or neighbors. Any Roofing Sub-Contractor you hire should have either a General Contractor's or home improvement license and proper liability and property damage insurance. They must be able to show you up-to-date proof of having this insurance. The Contractor should guarantee this work too. Get the guarantee in writing, spelling out exactly what will be covered.

Remember to get proof of insurance from your Sub-Contractors. If a Roofing Sub-Contractor falls off your roof they might own your house if you and they are not properly insured.

Soffit, Fascia, Down Spouts

Next to the roof, gutters and down spouts are your house's most important defense against damage from moisture and rot. Once the gutters and down spouts are put in place, check how the water will be directed away from your house and in the direction you desire it to travel.

Also, after it rains, inspect your crawl space in the attic with a bright light, looking for any water. This will indicate if there is a problem with your gutters and down spouts or a crushed drainage pipe.

Check and clean your gutters right after construction and at least once a year. One of the fastest ways to get water in your attic crawl space or basement is by not taking care of the drainage from your roof. You may be surprised how leaves and construction materials can accumulate in the gutters even before you have finished the house.

We decided to add gutter guards that snap on the gutters which allow rain to enter the gutter but stop any leaves and debris from entering the gutters. This saves having to clean or check the gutters more than once per year.

Before the Interior Walls are Covered...

Before you have your interior walls covered, check that these items are completed as applicable:

- sink drain rough-ins

- water supply for the sinks, showers, tubs

- cold water for the toilets

- exhaust system for the bathrooms and laundry room dryers

Here is the Master Bathroom showing electrical wiring,
water lines and drain lines.

Check that these items are completed as applicable - continued:

    - hot water for the dishwasher

    - hot and cold water for the clothes washer

    - gas lines

    - appliance vents

    - built-in vacuum system

    - electric wires and doorbell

This is the Master Bedroom Fireplace, electrical wiring,
including cable and phone lines.

Check that these items are completed as applicable - continued:

- intercom system

- phone lines

- alarm system

- heating and cooling ducts

- clothes chute

- cable lines

Our garage door systems and water and drain line locations.

Check that these items are completed as applicable - continued:

- extra support wood for closet racks to connect to

- recessed lighting in all the places you want it

> ## Tip
>
> Take the time to photograph and video tape all the electrical circuitry and drain and water lines to have record should you want to make any changes to something after the walls are all in place. We cannot tell you how many times we looked at the photos before pounding a nail into the wall just to make sure that we would not pound a nail into a drain pipe or water line.

Insulation

The Sub-Contractor we hired for drywall also installed insulation. We priced insulation at various R-values. To our surprise, this same Sub-Contractor gave us an estimate for materials and installation for insulating our whole house for just about what it was going to cost us to purchase the insulation ourselves. This was an easy decision to let the experts supply the materials and do the work.

When having your house insulated, do not forget to include areas other than just the outside walls. We sound-proofed our bedrooms (especially our Master Bedroom) with insulation on the walls and beneath the flooring. We insulated all our bathrooms and powder room to reduce any noise. Also consider insulating any water drain pipes which could be noisy as water passes through.

Drywall

Drywall was one of the pleasant surprises during construction. We were expecting a much higher estimate. We do not believe we could have saved any money by doing this step ourselves. This was one area where we knew it was best to let the experts install. We selected drywall over plastering since it is less likely to crack than plaster and it's easier to repair. Drywall must be painted because its finished appearance is not uniform.

The Drywall Contractor should supply a written estimate based on the number of sheets of drywall that must be hung and finished. The key to an excellent drywall job really begins with the carpenters, plumbers and electricians. If the framing is not straight and there is not proper backing in the corners or along the ceilings, the drywall hangers will have a hard time hanging the boards. You can help the process by checking for this during the framing stage. We did a lot of shaving and sanding before the drywall hanging was started.

Now Solomon was building his own house thirteen years, and he finished all his house
- 1Kings 7:1, NAS Study Bible

CHAPTER SEVEN

Once the Walls are Up

Interior Painting

Since the start-up costs of getting into the painting business are considerably less than some other trades, there are many first-time or part-time painters around. When you shop for a painter, it makes sense to look for someone who has been in the business for a few years and has earned a list of satisfied customers.

If you have a particular brand and model of paint you prefer, these should be specified in your contract. You can help minimize disappointments if you keep listing as many details as you expect to be included.

We rented a mechanical paint sprayer and did the work ourselves. If you do this, make sure you cover everything that you do not want painted. Spray paint will go everywhere. We liked how quickly the job was done and how evenly the paint was applied. We do admit that if the house will need to be re-painted in the future, we will hire professional painters. We would not have the opportunity to spray paint again like we did before the house was finished.

When you go to the Paint Supplier to determine how much paint you will need, make sure to say that you are spraying paint (if you in deed are spraying). If you spray the paint, you will use a lot more paint than if you just roll it on.

Finish Electrical

Now that the walls are up it is time to cover all those outlets, install those lights and fans, and check that everything is in good working order. You can purchase an outlet checker. This will check for proper grounding and hot and neutral connections.

Make sure you waterproof any tub and shower lights. These can easily be exposed to water. Keep safety in mind while working and installing things at home. Test all GFI circuits on a monthly basis. If water or outdoors is any part of the environment, use a GFI breaker or outlet for safety purposes. Contact your Local Inspector for GFI (Ground Fault Interrupt) regulations.

Flooring

Before laying any flooring down over the sub-floor, check to make sure all the nails or screws are placed at or below the surface level. Check for any waste or spills on the sub-floor such as glue, caulking or paint. Also check for any wood irregularities such as bumps, splinters, or grooves. You will want to remove any foreign matter, sand any bumps or splinters, and consider the size of the groove as to if it needs to be filled. Even small irregularities can show or be felt through your flooring or carpeting.

If you have used grout anywhere with tile in the house (kitchen, bathrooms, mud room, entranceway), make sure you seal it. This will help to keep water and dirt getting into the grout lines which will help the grout stay cleaner longer. Make sure you seal it only when the grout is very dry. We used a rose-colored grout for our tiles. We wanted to select a color that would not show dirt as quickly as a white grout. That is a common problem we have heard by many - that the white or very light colored grout gets dirty and is almost impossible to get it back to its original color.

Also, if anything sounds questionable or too good to be true from the salesperson, call the flooring manufacturer with your questions. When we were purchasing our floating hardwood flooring the salesperson told us we could not use a certain type of base wood. We found the salesperson to be incorrect after we spoke with the flooring manufacturer.

We decided to install a dark cherry colored wooded floor. It looks great when it is clean but is shows every piece of dust as it gets dusty!

Here's the entrance foyer and living room
hardwood floor under construction.

When we purchased the flooring the price was approximately $8,000. The Salesperson said that we would save the cost of the sales tax if we also had their Installers install the flooring. We asked for a quote. The quote was $10,000 to install the floor (more than the price of the flooring!). We decided to lay the floor ourselves. There were a couple of special tools required to lay the floor, but the price was right.

We rented and purchased several "How-to" videos for our home building project. These may also be available at your local library.

Tip

We found out that it costs as much as or even more to install ceramic tile or a hard-wood floor than to purchase it. Ask how much it costs to install before looking at any non-carpeted flooring. Also, sale prices do not typically reduce the installation price. Ask before signing!

Finish Heating and Air Conditioning

The wall and floor registers will need to be installed. This is something that is not too difficult to do where you can save some money. Do not forget to check and clean or change your furnace filters often when you first begin to use it. The house may be new, but the heating / cooling system collects lots of dust during construction, and sometimes the filter can be very dirty the day you move in.

Consider hiring a Duct Cleaning Company to come in and give your whole system a cleaning. They may be able to get to areas where you now cannot easily reach since everything is connected.

Cabinets

After pricing cabinets and comparing the quality and style for what we wanted, we decided to make our own. The first thing we did was purchase the numerous tools we would need for the job. This included getting a joiner, a shaper / router, planer / molder, a biscuit cutter long clamps and an air handler system to keep the dust down.

One thing we did learn when we went out obtaining estimates was that there are many excellent Kitchen Designers working for your local Home and Garden Centers. Do not hesitate to see what they have to offer. If you are not in the market to make your own cabinets, these can be reasonable depending upon what you are looking to purchase. We did pick up a few good ideas in talking with the Kitchen Designers.

If you do have cabinets installed, make sure you find out what they guarantee. Find out if you have the choice of cabinet fixtures (hinges and handles). And do they guarantee that your desired appliances will properly fit into the appliance openings (built-in ovens, drop-in stove tops, refrigerators, dishwashers, and odd-shaped sinks).

There are so many styles and prices. We went to a lot of new house "Open Houses" and looked at all the cabinetry. It is surprising of the poor quality that some very expensive homes have in place. You get what you pay for and also what you know to look out for. We would take photos of things that we liked or wanted to duplicate when we would visit these "Open Houses" too.

Counter Tops

We had changed our mind so many times with counter tops. We have looked at the reasonable laminate style tops, to the more expensive granite tops. Since cost was an item that we needed to consider, we elected to go with several types of counter tops. We like the ease of installing along with the inexpensive price tag of laminate. We decided to purchase the laminate overlay and add our own wood fronts for a "designer" look. The wood is stainable and is able to easily match the cabinets.

For our bathrooms, we selected the pre-made laminate counter tops which were very inexpensive, purchased at our local Home and Garden Center. We thought that we could change the color or design in the future without feeling we wasted much money. This will allow versatility in our bathrooms in years to come.

For our whirlpool tub landing in the Master Bath we selected our hardwood flooring. This matches the trim and cabinetry in our Master Bath.

To give you an example of cost, the price of a 10 foot piece of pre-cut laminate counter top with a back splash is approximately $60. For a 10 foot piece of granite counter top with a back splash, you might spend $1,250. When the price is quoted for granite, it may be per inch rather than per foot. Check the fine details when price comparing.

Finish Plumbing

This was an area where we were able to save money by purchasing the finish plumbing items ourselves. The problem we did run into was that we started upgrading these items, thinking of how much money we were saving by doing so much of this on our own. This turned into us going over our budget in this department. We selected "designer" colors for our Master Bath. Certain colors can almost double the cost. We also selected fancier fixtures to go with the "designer" colored bathroom items.

If you do have a Sub-Contractor complete the finish plumbing and they are purchasing the supplies, make sure they know all the measurements before they begin purchasing supplies. You may also want them to consult with you if anything is not available or is not within your price range before they make decisions for you.

This is an area that can easily go over your budget if you do not watch what is purchased. We were out looking at faucets one day and found a nice looking set with a price tag of $298. We thought the price was high. The Salesperson then told us that the price was for the faucet only and not the faucet valves and handles. Those would be another $110 each. We realized that the place we were in catered to folks who do not shop around.

Interior Trim

We saved interior trim (for floors, windows and crown molding) as one of the last items to complete. We found that after having an opportunity to look at the house, window casings, door casings, floor coverings and walls, we were able to determine what would look best with each room. Where we installed dark hard wood floors, we found that a white trim looked good since there was a dark floor already and our walls are painted white.

We installed dark (cherry) stained trim in the rooms where we have carpet. Our carpet is light colored and the dark contrast looks good for our taste. Our window and door casings are all dark (cherry) stained. We have our walls white and we like the contrast.

We did decide to install rosettes for all of the window and door casing corners. This makes connecting the trim much easier since we do not have to miter all the corners.

Tip

If you decide to install the interior trim yourself, seriously consider in investing in an automatic nailer. We selected a style that was cordless, powered by lightweight fuel cartridges. This made the trim installation a breeze and **one person could easily do the job**. Also purchase a good miter saw for making nice, clean cuts with the wood.

Pre-Hung Doors

When purchasing doors, it is much easier to install a hung. We did find that only a few "standard" sizes of p shelf. It gets more expensive when you have special size hung doors. You do have the benefit of selecting your own if you decide not to purchase pre-hung doors.

There were a few locations in our house that did not have standard door openings. Two of the locations were the entrance to our Sun Room and the entrance to our Den (this entrance was for a double-door). Instead of either adjusting the doorway to fit a smaller door or purchasing a more expensive special order door, we elected to make the doors ourselves. This allows you to add another personal touch to your house.

Staining

We elected to stain most everything rather than paint over the wood. We purchased oak for all of our trim and cherry for our cabinetry. The areas where we did paint were non-stainable floor trim and our fiberglass doors. The "secret" to good staining is the preparation work. You must sand the wood to a smooth finish if you want the stain to look its best.

There are sanding aids (automatic sanders and sanding bars) to help you in this process. We applied two coats on everything that we stained. We used a well- known stain purchased from our local Home and Garden Center. We first looked at samples they had posted and selected our color from there.

There are water-based stains available which make clean-up very easy. We decided to use a stain with mineral spirits. If you are going to stain a soft wood such as Pine, you might want to use a pre-stain conditioner which will allow the stain to cover evenly over the wood.

Door Knobs

We decided to spend a few dollars more and purchase the door levers rather than the standard round door knobs. We find that this makes opening a door much easier when our hands are full and unavailable to turn a door knob. However, doorknobs may be easier to "child-proof" than door levers.

We did use the regular door knob style for our doors to the outside elements. These door knobs were already provided with the package.

After all the Sub-Contractors are done, change the lock pins and keys to all the outside entries. You can purchase these lock pin kits from your local Hardware or Home Center. You will be surprised how many people will have keys to your house.

Carpeting

We found that carpeting comes typically in 12 foot and 15 foot width rolls. The Berber-style carpeting shows carpet seams much quicker than other types of carpeting. If your room has both dimensions wider than 15 feet, you cannot avoid a seam. You may want to consider this before purchasing something like Berber-style carpeting unless you have seen a carpet seam like this before and do not feel it will create an eyesore in your room.

When purchasing your carpeting, do not forget to ask:

- Does the purchase price include installation?

- Does installation include the price of the pad

- What size is the price for (square feet or square yards)?

- What is the quality of the carpet padding?

- How much extra will it be to upgrade the carpet padding?

If you are picking a carpet with a sculptured design, have the Salesperson pull out the carpet to see the full effect of the design. We first selected a sculptured design on the Berber Carpet but once we saw it rolled out, we decided not to go with any sculptured look.

Tip

A cushiony, higher quality carpet pad adds much more to the luxury of your carpet than the higher expense in the carpet. Test it out at the store with their carpet pad samples.

Required Inspections

Depending upon the area you move to or the bank where you obtain your construction loan, you can be required to various different inspections which could include:

- Rough Electrical Inspections

- Rough-Framing Inspection

- Permanent Power Connection Inspection

- Finished Electrical Inspection

- Plumbing Inspection

- Sewage Tap-In Inspection

- Gas Line Inspection

- Drain System Leak Test Inspection

- Footer Inspection

- Foundation Inspection

- Progress Inspections for your Mortgage Company

If inspections are not required, you may still want to hire various individuals for inspections. Sub-Contractors do take occasional "short-cuts" when they are busy. It could be worth the time and money to get various items inspected.

Insurance

Loan Institutions will typically require you to have your Home Insurance purchased up front when you go for your Construction Loan. Once you complete your home you may want to have the Insurance Agent come out and inspect your home for re-assessment purposes. You will want to make sure you are adequately covered.

We found that once we were done, our home was worth a bit more than what we originally estimated. This was partly due to a small amount of inflation, and largely due to many upgrades we incorporated along the building process. These add value to your house's worth.

Do not forget to follow up with your home insurance. It's an area where you will not want to be caught short when you need to use it.

Mine Subsidence Insurance

We built our house in an area where there is a lot of coal mines. We contacted our local Department of Environment Agency for a Coal Status Mine Report. Because our house would sit above a coal mine we were able to purchase Mine Subsidence Insurance. We purchased this from the Department of Environmental Protection (DEP). Insurance from the government can be obtained up to a value of $150,000. If the value of your house is greater than this, your local Agency can recommend what Insurance Company in your area will provide coverage for over this amount.

We found it necessary to obtain Mine Subsidence Insurance for over the government limit. We made the mistake of purchasing up to the limit from the government then seeking another insurance company for coverage over the $150,000. What we found was that the Insurance Company that provides coverage over the $150,000 limit will cover the entire amount. We did not need to purchase insurance from the government. We sought a refund after securing the Mine Subsidence Insurance from another Insurance Agent.

Driveway and Sidewalks - Concreted

We decided to install a cement driveway and stone sidewalks. We felt that cement will be cooler than asphalt in the hot summer. Our community required some type of paved driveway surface in our covenants. You may elect to use gravel initially. This can always be used as a base for cement or asphalt later. Make sure you budget for this if your community requires a paved driveway.

Also remember that once you have your driveway paved with cement or asphalt, you cannot (or should not) have large trucks using it. Plan the work so that things such as landscaping needs or someone suppling extra dirt, gravel, rocks, or cement for things such as your back yard, patio or sidewalks is done before the driveway is completed.

Landscaping

Landscapers go in and out of business frequently, so one indication of a good Landscaper is how long they have been in business. Guarantees from a company that goes out of business in the fall will do you little good next spring.

Call around to various Landscapers listed in the phone book (a good sign that they have been in business for awhile). Ask if they will give you a free estimate. This may save you time and money by finding this information out up front. Most good Landscapers will offer a free layout of plants, trees, shrubs and mulch fields.

The Landscaper that we selected would not plant grass seed until September. For our climate, they indicated that our yard would burn and not grow if we had it planted in the hotter summer months. Make sure to obtain the data sheets on how to properly tend to your landscape plantings. Some plants require a lot of maintenance.

Occupancy Permit

Once you are ready to move in (even if you are permitted to move in before all the construction is completed), go to your local community office, typically where you got your Building Permit, and obtain your Occupancy Permit. This sets everything in motion that you are now an official resident of that community and now have the pleasure of paying taxes.

Move In

Hurray! The day has come where you are moving into your wonderful new home. The one that you built. What a wonderful feeling to get that first night's rest in the fruit of all your hard labor. It's time to sit back, relax, and enjoy - at least until morning!

For every house is built by someone, but the builder of all things is God. - Hebrews 3:4, NAS Study Bible

Appendix 1

Sub-Contractor Bid Request Sheets

Here are several examples of Sub-Contractor Bid Request Sheet (Bid Sheets follow this summary list). Each Bid Sheet will list the pertinent type of information to request on each type of Bid Request. We have omitted names and address for simplicity of the examples.

Surveyors
Excavation / Removal
Footer Installation
Block Installation
Brick Installation
> Note: You may want the same Sub-Contractor to do the Footer, Block and Brick so care will be made to assure proper fit for these critical foundation elements. Our example Bid Sheet has these three combined.

Framing
Beams and Lintells Supply
Concrete Finishing
Roofing
Siding
Electric
Heating
Plumbing
Insulating
Gutters
Garage Doors

Request for Bid from the Surveyor Sub-Contractor

To: (Each Possible Surveyor Sub-Contractor)
Name of Contact Person:
Address:

Phone Number: **Fax Number:**
Best time to call:

From: (You)
         (Owner and General Contractor)
Address:

Phone Number: **Fax Number:**

Subject: Bid Request for Surveying Property

Work Required by Surveyor Sub-Contractor:
> Surveying a house located in Small Town, USA. The house will be located in Lot #45 on Happy Lane. Directions to the site are as follows: (provide directions to your plot of land).

The Surveyor Sub-Contractor will submit an itemized bid for the following work:
> Survey the above mentioned property. Supply the Owner / General Contractor with a Certificate of Survey.
>
> Locate and stake out the building footprint for the Excavator.

Sub-Contractor Insurance:
> The Sub-Contractor will verify Certificate of Insurance.

Materials supplied by Owner / General Contractor:
> The Owner will supply all the print materials required for surveying the desired model (including a plot plan of the property and a copy of the house blue prints).

Request for Bid from the Surveyor Sub-Contractor
Page 2

Materials supplied by Sub-Contractor:
> The Surveying Contractor will set the property corners; prepare the plot plan for the building permit; stakeout the house for excavator on offsets along the side property lines; will survey the final building location (as-built); and, will prepare a certified plot plan to be used for occupancy, mortgage company, etc...

What needs to be completed before the Surveyor Sub-Contractor arrives:
> The Building Permit will be secured by the Owner

Lead Time:

Completion Time (days to complete):

Special Items: (attach with Bid Sheet):
> The plan drawings will be provided for the actual survey. Restrictions for this housing plan include keeping the building placement at least 20 feet from the adjoining lots and that the house is placed 35 feet from the street.

> Note that two of the side rooms contain bay windows; that the garage is on the side; and, there is a four foot extension included on the kitchen.

Please submit your bid in writing within 2 weeks of receiving this request of bid to the above address or fax.

Request for Bid from the Excavation Sub-Contractor

To: (Each Possible Excavation Sub-Contractor)
Name of Contact Person:
Address:

Phone Number: Fax Number:
     Best time to call:

From: (You)
        (Owner and General Contractor)
Address:

Phone Number: Fax Number:

Subject: Bid Request for Excavating the Lot and Hauling Excess Dirt

Work Required by the Excavating Sub-Contractor:
     Excavating the lot and hauling away any excess dirt for a house located
     in Small Town, USA. The house will be located in Lot #45 on Happy
     Lane. Directions to the site are as follows: (provide directions to your
     plot of land).

The Excavation Sub-Contractor will submit an itemized bid for the following
work:
     Excavate the above mentioned property for the buried foundation;
     spread sand in excavated area; provide trenches for the utilities (water,
     sewage, electric, phone and cable); clear two 40' x 8' areas for the
     framing packages; spread slag down to allow the tractor trailers to back
     onto the property; back-fill around the foundation at a later time; and,
     haul away any stumps, tree pieces, and soil (not needed for back-fill).

Subcontractor Insurance:
     The Subcontractor will verify Certificate of Insurance.

Request for Bid from the Excavation Sub-Contractor
Page 2

Materials supplied by the Excavation Sub-Contractor:
    The Excavating Sub-Contractor will supply the necessary equipment to excavate the land and haul the excess dirt, rock, trees, and stumps away from the property site.

What needs to be completed before the Excavation Sub-Contractor arrives:
    The Building Permit and the required Certificate of Survey will be secured by the Owner. The Stake-Out will also be complete.

Lead Time:

Completion Time:

Special Items:
    This excavation is for the attached drawings with the following options:

- Side entry entrance to the Basement
- 3 car Garage - side entry (1 single and 1 double door)
- Garage floor will be approximately 3 feet lower than the Kitchen elevation
- Sun Room
- 4 feet extension Kitchen option
- 13 courses of block
- Bay window in the Living Room (North side)
- Bay window in the Study (North side)

    The property has a slight upward grade (approximately a 7 foot rise to a 31 foot run) to a flat area. There is a high probability that there will be rock approximately 6 feet down and could require special jack hammering equipment to remove the rock.

Please submit your bid in writing within 2 weeks of receiving this request of bid to the above address or fax.

Request for Bid from the Footer, Block & Brick Sub-Contractor

To: (Each Possible Footer, Block & Brick Sub-Contractor)
          Note: You may elect to make these all separate Bids
Name of Contact Person:
Address:

Phone Number: **Fax Number:**
          Best time to call:

From: (You)
          (Owner and General Contractor)
Address:

Phone Number: **Fax Number:**

Subject: Bid Request for laying the Footer, Block and Bricking the House

Work Required by the Footer, Block & Brick Sub-Contractor:
          To provide the following labor for the Owner / Builder Home, Model XYZ:
-         To frame and pour a concrete footer for a basement and a 3 car garage
-         Lay a block foundation
-         Sand the footing to reduce falling mortar build-up
-         Brick four sides of the house
          Note: The rear side will be bricked with the exception of the extended gas fireplace. Also, the two bay windows on the North side of the house will have siding installed.
-         Adding decorative brick work at all corners
-         Adding decorative brick work around all windows to eliminate the need for shutters
-         Placing brick work above the front door to eliminate the need for the wood work in the original house design
-         Placing the purchased house number on the front of the house near the front doorway

Request for Bid from the Footer, Block & Brick Sub-Contractor
Page 2

The Footer, Block & Brick Sub-Contractor will submit an itemized bid for the following work:

> Install the footer and block on the above mentioned property followed by bricking the house once the framing is complete. All is defined in the Work Required listed above.

Sub-Contractor Insurance:
> The Sub-Contractor will verify Certificate of Insurance.

Materials supplied by the Footer, Block & Brick Sub-Contractor:

- All concrete, lumber, rebar, gravel, water, as well as all tools required to complete the footer work.
- All cement blocks, mortar, sand, water, as well as tools required to complete the foundation wall.
- All mortar, sand, water as well as tools, scaffolding, and mixing equipment required to brick all four sides of the home.

Materials supplied by the Owner / Builder General Contractor:

> The Builder will supply all the prints / drawings showing the block foundation walls for the Owner / Builder Home, Model XYZ. The Builder will also provide the Brick.

What needs to be completed before the Footer, Block & Brick Subcontractor arrives:

> The Excavation work must be completed.

Request for Bid from the Footer, Block & Brick Sub-Contractor
Page 3

Lead Time:

Completion Time:

Special Items:

> **The Owner / Builder Home, Model XYZ will have the following options:**
> - Side entry entrance to the Basement
> - 3 car Garage - side entry (1 single and 1 double door)
> - Garage floor will be approximately 3 feet lower than the Kitchen elevation
> - Sun Room
> - 4 feet extension Kitchen option
> - 13 courses of block
> - Bay window in the Living Room (North side)
> - Bay window in the Study (North side)
> - Bricked on all four sides

Please submit your bid in writing within 2 weeks of receiving this request of bid to the above address or fax.

Request for Bid from the Framing Sub-Contractor

To: (Each Possible Framing Sub-Contractor)
Name of Contact Person:
Address:

Phone Number: **Fax Number:**
    Best time to call:

From: (You)
        (Owner and General Contractor)
Address:

Phone Number: **Fax Number:**

Subject: Bid Request for framing the Owner / Builder Home, Model XYZ

Work Required by the Framing Sub-Contractor:
    To provide the following labor for the Owner / Builder Home, Model
    XYZ:
-    Break down the framing package
-    Verify that the foundation meets the required dimensions as per
     the plans for the Owner / Builder Home, Model XYZ
-    Set the bearing beams
-    Check and set the stanchion posts
-    Apply sill sealer and sill plate, joists, and bridging
-    Glue joists and attach deck tongue and groove plywood to joists
-    Erect gable walls and brace
-    Truss placement
-    Plywood roof placement, soffit, fascia, ridge vents, and roof felted
-    Seal and wrap all joints and corners, as well as any over hangs
     with asphalt impregnated paper
-    Repair damaged areas to the exterior sheathing
-    Interior panel placement
-    Outside trim, windows, sills, and exterior doors installed
-    Stairs set
-    Install baffles between trusses

Request for Bid from the Framing Sub-Contractor
Page 2

The Framing Sub-Contractor will submit an itemized bid for the following work:

> Framing the above mentioned property as is defined in the Work Required listed above.

Sub-Contractor Insurance:

> The Sub-Contractor will verify Certificate of Insurance.

Materials supplied by the Framing Sub-Contractor:

> The Framing Sub-Contractor will be required, as a minimum, to supply the equipment necessary for framing the described Home above.

Materials supplied by the Owner / Builder General Contractor:

> The Builder will supply all the materials required for framing the Owner / Builder Home, Model XYZ with the exception of items mention above that are to be supplied by the Framing Sub-Contractor.

What needs to be completed before the Framing Sub-Contractor arrives:

> The excavation, footer, foundation, insulation, backfilled excavation around the foundation, and the french drains completed.

Lead Time:

Completion Time:

Request for Bid from the Framing Sub-Contractor
Page 3

Special Items:

The Owner / Builder Home, Model XYZ will have the following options:

- Side entry entrance to the Basement
- 3 car Garage - side entry (1 single and 1 double door)
- Garage floor will be approximately 3 feet lower than the Kitchen elevation
- ABC Brand Windows
- Draft vent gas fireplace in the Family Room
- Large arched Window above the fireplace in the Family Room
- Sun Room
- Skylights in the Sun Room
- 4 feet extension Kitchen option
- 13 courses of block
- Bay window in the Living Room (North side)
- Bay window in the Study (North side)
- Double doors in the Master Bedroom
- Double doors in the Master Bathroom
- Tray ceiling in the Master Bedroom
- Draft vent gas fireplace in the Master Bedroom
- Draft vent gas fireplace in the Master Bathroom

Please submit your bid in writing within 2 weeks of receiving this request of bid to the above address or fax.

Request for Bid from the Beams and Lintells Supplier

To: (Each Possible Beams and Lintells Supplier)
Name of Contact Person:
Address:

Phone Number: Fax Number:
        Best time to call:

From: (You)
        (Owner and General Contractor)
Address:

Phone Number: Fax Number:

Subject: Bid Request for Beams and Lintells for the Owner / Builder Home,
                Model XYZ

Materials supplied by the Beams and Lintells Vendor:
        The vendor will supply the following materials:
        -       1 Beam W8 x 18 x 32'8"
        -       1 Beam W8 x 18 x 20'8"
        -       1 Beam W12 x 26 x 20'
        -       1 Lintell 17' for garage door
        -       1 Lintell 9' for garage door
        -       2 Lintells 9' for bay windows
        -       6 Lintells 34" for sun room casement windows
        -       5 Lintells 3'4" for Master bedroom

Materials supplied by the Owner / Builder General Contractor:
        The Builder will supply all the prints / drawings showing the beams and
        lintells for the Owner / Builder Home, Model XYZ.

What needs to be completed before the Concrete Subcontractor arrives:
        The Excavation, footer, foundation, framing and plumbing rough-in
        work must be completed.

Request for Bid from the Beams and Lintells Supplier
Page 2

What needs to be completed before the Concrete Sub-Contractor arrives:

    The Excavation, footer and foundation

Lead Time:

Completion Time:

Special Items:

    The Owner / Builder Home, Model XYZ will have the following options:

-     **Side entry entrance to the Basement (single door access)**
-     **3 car Garage - side entry (1 single and 1 double door)**
-     **Garage floor will be approximately 3 feet lower than the Kitchen elevation**
-     **Sun Room**
-     **4 feet extension Kitchen option**
-     **13 courses of block**
-     **Bay window in the Living Room**
-     **Bay window in the Study**

Please submit your bid in writing within 2 weeks of receiving this request of bid to the above address or fax.

Request for Bid from the Concrete Sub-Contractor

To: (Each Possible Concrete Sub-Contractor)
Name of Contact Person:
Address:

Phone Number: **Fax Number:**
        Best time to call:

From: (You)
        (Owner and General Contractor)
Address:

Phone Number: **Fax Number:**

Subject: Bid Request for labor and materials on concrete finishing the
        Owner / Builder Home, Model XYZ

Work Required by the Concrete Sub-Contractor:
        To provide the following labor for pouring and finishing the Owner /
        Builder Home, Model XYZ:
        -    Basement
        -    3 Car Garage
        -    Driveway (approximately 25' wide by 53' long)
        -    Front Porch and steps for the front porch
        -    A walkway (approximately 3' wide by 42' long)
        -    Basement side entry and steps for the side entry

The Concrete Sub-Contractor will submit an itemized bid for the following
work:
        Cementing the above mentioned property as is defined in the Work
        Required listed above.

Subcontractor Insurance:
        The Subcontractor will verify Certificate of Insurance.

Request for Bid from the Concrete Sub-Contractor
Page 2

Materials supplied by the Concrete Sub-Contractor:
All concrete, mortar, sand, gravel, water, as well as all tools required to complete the concrete work.

Materials supplied by the Owner / Builder General Contractor:
The Builder will supply all the prints / drawings showing the areas of the basement, garage floor, porch, walkway, driveway and steps for the Owner / Builder Home, Model XYZ.

What needs to be completed before the Concrete Sub-Contractor arrives:

The Excavation, footer, foundation, framing and plumbing rough-in work must be completed.

Lead Time:

Completion Time:

Special Items:

The Owner / Builder Home, Model XYZ will have the following options:
-       Side entry entrance to the Basement (single door access)
-       3 car Garage - side entry (1 single and 1 double door)
-       Garage floor will be approximately 3 feet lower than the Kitchen elevation
-       Sun Room
-       4 feet extension Kitchen option
-       The house will be bricked on all four (4) sides

Please submit your bid in writing within 2 weeks of receiving this request of bid to the above address or fax.

Request for Bid from the Roofing Sub-Contractor

To: (Each Possible Roofing Sub-Contractor)
Name of Contact Person:
Address:

Phone Number: Fax Number:
        Best time to call:

From: (You)
        (Owner and General Contractor)
Address:

Phone Number: Fax Number:

Subject: Bid Request for labor and materials on roofing the Owner /
            Builder Home, Model XYZ

Work Required by the Roofing Sub-Contractor:
        To provide the following labor for installing shingles and shingle
        underlayment for ice damage protection on the following areas of the
        Owner / Builder Home, Model XYZ:
        -       Shingles on all roofs and bay windows
        -       Shingle underlayment on eave edges, hips, skylights and valleys

The Roofing Sub-Contractor will submit an itemized bid for the following
work:
        Roofing the above mentioned property as is defined in the Work
        Required listed above.

Subcontractor Insurance:
        The Subcontractor will verify Certificate of Insurance.

Request for Bid from the Roofing Sub-Contractor
Page 2

Materials supplied by the Roofing Sub-Contractor:
Underlayment material 60 to 80 mils, flashing for roof tie-ins as well as all tools required to complete the roofing work such as ladders and scaffolding.

Materials supplied by the Owner / Builder General Contractor:
The Builder will supply all the prints / drawings showing the areas for roofing the Owner / Builder Home, Model XYZ.

What needs to be completed before the Roofing Sub-Contractor arrives:
The Excavation, footer, foundation, foundation insulation, backfilled excavation around the foundation, French Drains, and framing complete along with the vent stacks cut through the roof.

Lead Time:

Completion Time:

Special Items:
The Owner / Builder Home, Model XYZ will have the following options:
- 3 car Garage - side entry (1 single and 1 double door)
- ABC Brand windows
- Draft vent gas fireplace in the Family Room
- Large arched window above the fireplace
- Sun room
- Skylights in the Sun room
- Bay window in the living room (side of house)
- Bay window in Study
- 4 feet extension Kitchen option

Please submit your bid in writing within 2 weeks of receiving this request of bid to the above address or fax.

Request for Bid from the Siding Sub-Contractor

To: (Each Possible Siding Sub-Contractor)
Name of Contact Person:
Address:

Phone Number: **Fax Number:**
    Best time to call:

From: (You)
    (Owner and General Contractor)
Address:

Phone Number: **Fax Number:**

Subject: Bid Request for labor and materials on siding the Owner / Builder Home, Model XYZ

Work Required by the Siding Sub-Contractor:
    To provide the following labor for installing siding on the following areas of the Owner / Builder Home, Model XYZ:
    -        the Sunroom - inside wall only
    -        the rear wall behind the Family Room
    -        the rear of the 4' extension on the Kitchen

The Siding Sub-Contractor will submit an itemized bid for the following work:

    Siding the above mentioned property as is defined in the Work Required listed above.

Subcontractor Insurance:
    The Subcontractor will verify Certificate of Insurance.

Request for Bid from the Siding Sub-Contractor
Page 2

Materials supplied by the Siding Sub-Contractor:
   All siding nails and chalking and whatever tools required such as ladders and scaffolding.

Materials supplied by the Owner / Builder General Contractor:
   The Builder will supply all the prints / drawings showing the areas for roofing the Owner / Builder Home, Model XYZ.

What needs to be completed before the Siding Sub-Contractor arrives:
   The Excavation, footer, foundation, foundation insulation, backfilled excavation around the foundation, French Drains, and framing complete along with the vent stacks cut through the roof and house bricked on three sides.

Lead Time:
Completion Time:

Special Items:
   The Owner / Builder Home, Model XYZ will have the following options:
   -       3 car Garage - side entry (1 single and 1 double door)
   -       ABC Brand windows
   -       Draft vent gas fireplace in the Family Room
   -       Large arched window above the fireplace
   -       Sun room
   -       Skylights in the Sunroom
   -       Bay window in the living room (side of house)
   -       Bay window in Study
   -       4 feet extension Kitchen option

Please submit your bid in writing within 2 weeks of receiving this request of bid to the above address or fax.

Request for Bid from the Electric Sub-Contractor

To: (Each Possible Electric Sub-Contractor)
Name of Contact Person:
Address:

Phone Number: Fax Number:
        Best time to call:

From: (You)
        (Owner and General Contractor)
Address:

Phone Number: Fax Number:

Subject: Bid Request for labor and materials on installing the necessary
              electrical wiring for the Owner / Builder Home, Model XYZ

Work Required by the Electric Sub-Contractor:
        To provide the labor for installing all the electrical wiring per the
        electrical drawings for the Owner / Builder Home, Model XYZ.

The Electric Sub-Contractor will submit an itemized bid for the following
work:
        Installing the electrical wiring for the above mentioned property as is
        defined in the Work Required listed above.

Sub-Contractor Insurance:
        The Subcontractor will verify Certificate of Insurance.

Materials supplied by the Electric Subcontractor:
        All wiring, siding nails and chalking and whatever tools required such
        as ladders and scaffolding.

Request for Bid from the Electric Sub-Contractor
Page 2

Materials supplied by the Owner / Builder General Contractor:
   The Builder will supply all the prints / drawings showing the areas for wiring the Owner / Builder Home, Model XYZ.

What needs to be completed before the Electric Sub-Contractor arrives:
   The Excavation, footer, foundation, foundation insulation, backfilled excavation around the foundation, French Drains, and framing complete along with the vent stacks cut through the roof and house bricked and sided.

Lead Time:
Completion Time:

Special Items:

   The Owner / Builder Home, Model XYZ will have the following options:
   -       3 car Garage - side entry (1 single and 1 double door)
   -       ABC Brand windows
   -       Draft vent gas fireplace in the Family Room
   -       Large arched window above the fireplace
   -       Sun room
   -       Skylights in the Sunroom
   -       Bay window in the living room (side of house)
   -       Bay window in Study
   -       4 feet extension Kitchen option

Please submit your bid in writing within 2 weeks of receiving this request of bid to the above address or fax.

Request for Bid from the Heating Sub-Contractor

To: (Each Possible Heating Sub-Contractor)
Name of Contact Person:
Address:

Phone Number: Fax Number:
        Best time to call:

From: (You)
        (Owner and General Contractor)
Address:

Phone Number: Fax Number:

Subject: Bid Request for labor and materials on installing the necessary
            Heating for the Owner / Builder Home, Model XYZ

Work Required by the Heating Sub-Contractor:
        To provide the labor for installing all the heating and air conditioning
        units and air supply and air return ducts per the drawings for the
        Owner / Builder Home, Model XYZ.

The Heating Sub-Contractor will submit an itemized bid for the following
work:
        Installing the heating and air conditioning system for the above
        mentioned property as is defined in the Work Required listed above.

Subcontractor Insurance:
        The Subcontractor will verify Certificate of Insurance.

Materials supplied by the Heating Subcontractor:

        All nails, screws, duct work, vents and whatever tools required such as
        ladders and scaffolding.

Request for Bid from the Heating Sub-Contractor
Page 2

Materials supplied by the Owner / Builder General Contractor:
The Builder will supply all the prints / drawings showing the areas for heating and air conditioning the Owner / Builder Home, Model XYZ. The heating and air conditioning units will be purchased separately and supplied to the building site.

What needs to be completed before the Heating Sub-Contractor arrives:
The Excavation, footer, foundation, foundation insulation, backfilled excavation around the foundation, French Drains, and framing complete along with the vent stacks cut through the roof and house bricked and sided.

Lead Time:
Completion Time:

Special Items:

The Owner / Builder Home, Model XYZ will have the following options:
- Side entry entrance to the basement
- 3 car Garage - side entry (1 single and 1 double door)
- Garage floor will be approximately 3' lower than the kitchen elevation
- Sunroom
- Bay window in the living room (side of house)
- Bay window in Study
- 4 feet extension Kitchen option
- 13 courses of block

Please submit your bid in writing within 2 weeks of receiving this request of bid to the above address or fax.

Request for Bid from the Plumbing Sub-Contractor

To: (Each Possible Plumbing Sub-Contractor)
Name of Contact Person:
Address:

Phone Number: Fax Number:
        Best time to call:

From: (You)
        (Owner and General Contractor)
Address:

Phone Number: Fax Number:

Subject: Bid Request for labor and materials on plumbing the Owner /
             Builder Home, Model XYZ

Work Required by the Plumbing Sub-Contractor:
        To provide the following labor for both rough and finish plumbing for
        the following areas of the Owner / Builder Home, Model XYZ:
        -       install 4" plastic drain pipe around the footer for drainage
        -       install drain pipes in the basement and garage areas
        -       stack through the roof hole cut
        -       install PVC drain, waste and vent system
        -       sewage tap-in
        -       install sewage and water supply lines into the home under the
                foundation
        -       install internal supply lines
        -       install tub surround for the main second floor bath
        -       install whirlpool tub in the master bath

The Siding Sub-Contractor will submit an itemized bid for the following work:
        Plumbing the above mentioned property as is defined in the Work
        Required listed above.

Request for Bid from the Plumbing Sub-Contractor
Page 2

Sub-Contractor Insurance:
   The Sub-Contractor will verify Certificate of Insurance.

Materials supplied by the Plumbing Sub-Contractor:
   All copper, PVC pipes, connectors, traps, toilet floor flanges, glues, solder, gas torches, and other such tools required to complete the plumbing requirements of this style home.

Materials supplied by the Owner / Builder General Contractor:
   The Builder will supply all the prints / drawings showing the areas for plumbing the Owner / Builder Home, Model XYZ as well as the fixtures, hot water tank, tub surrounds, whirlpool tub, toilets, sinks and faucets.

What needs to be completed before the Siding Sub-Contractor arrives:

   The Excavation and footer.

Lead Time:

Completion Time:

Request for Bid from the Plumbing Sub-Contractor
Page 3

Special Items:

The Owner / Builder Home, Model XYZ will have the following plumbing areas:

- rough-in in the basement for a Powder Room
- hot and cold supply for basement laundry
- 1" cold water manifold in the basement to feed the water needs of the home
- cold water outlets for each side of the house with internal water cutoff for each spicket
- hot and cold water outlets in the garage area
- first floor Powder Room (toilet and sink)
- second floor Main Bathroom with a tub surround, sink and toilet
- Master Bathroom glass shower stall, whirlpool tub, toilet and two sinks
- Kitchen will have two sinks - one on the main counter and one smaller one in the island
- hot and cold supply for the first floor laundry

Please submit your bid in writing within 2 weeks of receiving this request of bid to the above address or fax.

Request for Bid from the Insulating Sub-Contractor

To: (Each Possible Insulating Sub-Contractor)
Name of Contact Person:
Address:

Phone Number: Fax Number:
      Best time to call:

From: (You)
         (Owner and General Contractor)
Address:

Phone Number: Fax Number:

Subject: Bid Request for labor and materials on insulating the Owner /
             Builder Home, Model XYZ

Work Required by the Insulating Sub-Contractor:
      To provide the following labor for insulating the following areas of the
      Owner / Builder Home, Model XYZ:
      -       insulation of the foundation with extruded polystyrene
      -       insulation of ceilings and walls
      -       insulation of the basement
      -       foam insulation board for behind the bathroom tubs

The Insulating Sub-Contractor will submit an itemized bid for the following
work:
      Roofing the above mentioned property as is defined in the Work
      Required listed above.

Subcontractor Insurance:
      The Subcontractor will verify Certificate of Insurance.

Request for Bid from the Insulating Sub-Contractor
Page 2

Materials supplied by the Insulating Sub-Contractor:
Insulation, insulation foam boards, as all tools required to complete the insulating work such as ladders and scaffolding.

Materials supplied by the Owner / Builder General Contractor:
The Builder will supply all the prints / drawings showing the areas for insulating the Owner / Builder Home, Model XYZ.

What needs to be completed before the Insulating Sub-Contractor arrives:
Building Permit, survey and stake out completed, Excavation, footer, and foundation.

Lead Time:
Completion Time:

Special Items:
The Owner / Builder Home, Model XYZ will have the following options:
- 3 car Garage - side entry (1 single and 1 double door)
- Side entry entrance to the basement
- Garage floor will be approximately 3' lower than the Kitchen elevation
- Sun room
- 4 feet extension Kitchen option
- Bay window in the living room (side of house)
- Bay window in Study
- 13 courses of block

Please submit your bid in writing within 2 weeks of receiving this request of bid to the above address or fax.

Request for Bid from the Gutters & Down Spouts Sub-Contractor

To: (Each Possible Gutters & Down Spouts Sub-Contractor)
Name of Contact Person:
Address:

Phone Number: Fax Number:
        Best time to call:

From: (You)
        (Owner and General Contractor)
Address:

Phone Number: Fax Number:

Subject: Bid Request for labor and materials on installing gutters and
            down spouts on the Owner / Builder Home, Model XYZ

Work Required by the Gutters & Down Spouts Sub-Contractor:

        To provide the labor for installing the gutters and down spouts of the
        Owner / Builder Home, Model XYZ.

The Gutters & Down Spouts Sub-Contractor will submit an itemized bid for
the following work:
        Installing the gutters and down spouts for the above mentioned property
        as is defined in the Work Required listed above.

Subcontractor Insurance:
        The Subcontractor will verify Certificate of Insurance.

Request for Bid from the Gutters & Down Spouts Sub-Contractor
Page 2

Materials supplied by the Gutters & Down Spouts Sub-Contractor:
   All material required to install aluminum gutters and three first floor and 3 second floor down spouts, as well as all tools required to complete the installation of this work such as ladders and scaffolding.

Materials supplied by the Owner / Builder General Contractor:
   The Builder will supply all the prints / drawings showing the areas for installing the gutters and down spouts for the Owner / Builder Home, Model XYZ.

What needs to be completed before the Insulating Sub-Contractor arrives:
   Excavation, footer, foundation, foundation insulation, backfilled excavation around the foundation, French Drains, framing, exterior bricked, sided, and roof shingled.

Lead Time:

Completion Time:

Special Items:
   The Owner / Builder Home, Model XYZ will have the following options:
   -   3 car Garage - side entry (1 single and 1 double door)
   -   Side entry entrance to the basement
   -   Garage floor will be approximately 3' lower than the Kitchen elevation
   -   Sun room
   -   4 feet extension Kitchen option
   -   Bay window in the living room (side of house)
   -   Bay window in Study

Please submit your bid in writing within 2 weeks of receiving this request of bid to the above address or fax.

Request for Bid from the Garage Door Sub-Contractor

To: (Each Possible Garage Door Sub-Contractor)
Name of Contact Person:
Address:

Phone Number: Fax Number:
     Best time to call:

From: (You)
        (Owner and General Contractor)
Address:

Phone Number: Fax Number:

Subject: Bid Request for labor and materials on installing the Garage Door
             for the Owner / Builder Home, Model XYZ

Work Required by the Garage Door Sub-Contractor:
        To provide the labor for installing the garage doors (8' x 8' and 16' x 8')
        and electric openers per the drawings for the Owner / Builder Home,
        Model XYZ.

The Garage Door Sub-Contractor will submit an itemized bid for the following
work:

        Installing the two garage doors and automatic openers for the above
        mentioned property as is defined in the Work Required listed above.

Subcontractor Insurance:
        The Subcontractor will verify Certificate of Insurance.

Request for Bid from the Garage Door Sub-Contractor
Page 2

Materials supplied by the Garage Door Sub-Contractor:
- Two garage doors (one 8' x 8' and one 16' x 8')
- garage door openers for the 8' high doors
- all mounting hardware, tracks, locks, exterior opener switch for both doors

Materials supplied by the Owner / Builder General Contractor:
The Builder will supply all the prints / drawings showing the openings to receive garage doors for the Owner / Builder Home, Model XYZ.

What needs to be completed before the Garage Door Sub-Contractor arrives:
The Excavation, footer, foundation, foundation insulation, backfilled excavation around the foundation, French Drains, framing complete along with the garage floor laid and exterior bricked.

Lead Time:

Completion Time:

Special Items:

The Owner / Builder Home, Model XYZ will have the following options:
- Side entry entrance to the basement
- 3 car Garage - side entry (1 single and 1 double door)
- Garage floor will be approximately 3' lower than the kitchen elevation

Please submit your bid in writing within 2 weeks of receiving this request of bid to the above address or fax.

Appendix 2

A Step-by-Step Process of Home Building

Here's a **Sequence Chart** of when you may want to start each item:

Item	1st	2nd	3rd	4th	5th	6th	7th	8th	9th	10th	11th	12th	13th	14th	15th	16th	17th	18th	19th
Survey Land	X																		
Permits	X																		
Clear Land		X																	
Excavation of Foundation			X																
Lay Footer				X															
Foundation Block Installed					X														
French Drain					X														
Waterproof the Wall						X													
Install support Beams						X													
Framing							X												
Install Sewage Lines in Basement							X												
Concrete Basement / Garage							X												
Begin Plumbing							X												
Begin Electrical							X												
Begin Heating / Air Conditioning							X												
Roof Shingled								X											
Lay Bricks									X										
Install Siding									X										
Install Soffit, Fascia & Down Spouts										X									
Insulation										X									
Drywall											X								
Interior Painting												X							

Step-by-Step Process of Home Building Sequence Chart continued...

Item	1st	2nd	3rd	4th	5th	6th	7th	8th	9th	10th	11th	12th	13th	14th	15th	16th	17th	18th	19th
Finish Electrical													X						
Install Flooring													X						
Finish Heating / Air Conditioning													X						
Install Cabinets													X						
Install Countertops														X					
Finish Plumbing															X				
Install Interior Trim															X				
Install Inside Doors																X			
Stain																X			
Install Carpeting																	X		
Outside Driveway / Sidewalks																	X		
Landscaping																	X		
Obtain necessary Inspections																		X	
Obtain Occupancy Permit																		X	
Move In!																			X

Appendix 3
Glossary

Access Door	Door installed in a location permitting ability to enter into a place such as a basement or garage.
Acreage	An area defined in acres or portion of an acre.
Amenities	The attractiveness and value of real estate that conduces to material comfort or convenience
Ash Dump	Fireplace ash disposal opening
Asphalt	A residue from petroleum Can be used for paving driveways and waterproofing roofs and exterior walls
Assessment	A charge against real property made by a branch of government to cover the proportionate cost of an improvement such as a street, sewer or school
Assets	Items on a balance sheet showing the book value of property owned (subject to the payment of any debts).
Attached Garage	Garage that is not considered an integral part of the house but rather separately attached
Automatic Nailer	Tool for nailing which utilizes a power source such as electricity or a propellant.
Back Fill	Excavated or imported material used around the foundation or in filling a trench.
Bay Style Windows	A window (or set of windows) projecting outward from the wall of the house.
Beam	A horizontal supporting structural member made of wood or metal.
Bearing Beams	A beam that supports a part of the floors and roof above it.
Better Business Bureau	Organization whose mission is to promote and foster the highest ethical relationship between businesses and the public through voluntary self-regulation, consumer and business education, and service excellence.
Bid	A statement of what one will give or take for something such as an offer of a price to perform work.
Biscuit Cutter	Instrument used to cut slots in wood to join together with a thin oval-shaped wafer of compressed wood known as a Biscuit.
Blacktop Driveway	An asphalt-type covered driveway.
Block	A hollow rectangular building unit constructed of cement or other type of material.
Blueprint	A photographic print of the architects' plans in white on a bright blue ground.
Board	A group of persons having managerial, supervisory, or investigatory powers (Board of Directors)
Bondable	The Sub-Contractor is insured, having an agreement pledging surety for financial loss caused to another by the act or default of a third person or by some contingency by which the third person may have no control.
Breaker	Electrical circuit breaker
Breakfast Nook	A secluded, recessed place in or near the kitchen area typically used for the morning meal
Bricklayer	One who lays bricks.
Bridging	Small pieces of wood or metal used to brace floor joists
Building Codes	Regulations established by state and local governments stating the minimum standards for construction.

Building Inspector	Person who inspects various phases of the construction process to ensure it meets proper building code restrictions
Building Line	A line fixed at a certain distance from the front of a lot, beyond which no structure can project.
Building Paper	A general term for papers, felts and similar sheet materials used in buildings.
Building Permit	A written warrant or license granted by one having authority for another to construct or add to an existing construction.
Building Restrictions	There is a fixed line at a certain distance from the front and / or sides of the lot which marks the boundary of the area within no part of any building may project.
Casement Window	A window that swings out to the side on hinges.
Casing	A molding used to trim door and window frames.
Cathedral Style Ceiling	A raised, pitched ceiling for the effect of having a large, spacious room
Caulking	Putty substance for sealing joints.
Cement	A fine powder that produces a bonding paste when mixed with water.
Central Vacuum System	Also known as whole house vacuum system. The vacuum component is hidden and piping is installed all over the house with outlets / access in every necessary location
Central Air Conditioning System	Also known as whole house air conditioning. One main unit is set up to cover the whole house Cool air is circulated through the duct work
Certificate of Insurance	A written document signed by a responsible representative of the insurance company and stating the exact coverage and period of time for which the coverage is applicable in accordance with requirements of the contract documents.
Clearing the Property	Removal of trees, weeds and trash from the property so that construction can begin.
Climate / Timer Controller	Instrument used to control the inside temperature by either setting the temperature or by setting a timer to adjust the temperature at various times of the day
Closing	The process by which the promises and agreements between the parties to a real estate transaction are fulfilled.
Clout	Influence or pull with someone.
Coal Status Mine Report	Government report of the status of mines beneath the property It should include the depth of the mine from property surface and the last activity of the mine.
Column	A perpendicular supporting member supporting loads
Completion Times	Days or dates listed when work will be completed from the state date.
Concrete	A construction material made of cement, fine aggregate or sand, coarse aggregate or gravel, and water mixed together.
Concrete Block	A hollow rectangular building unit constructed of cement or other type of material
Construction Loan	Loans for home building during the construction phase. Loan terms are typically 6 months to 2 years then switch over to a convention style mortgage.
Contract	A legally enforceable agreement between competent parties to do or not do certain things for consideration
Contractor Insurance	Insurance to protect the Contractor from claims which may arise from the Contractor's operations, including any employees of the Contractor
Contractor	A professional who specializes in one or more disciplines of construction such as excavation, plumbing, etc. He / She may be hired to provide labor only or labor and materials
Contractor's Loan	Loan provided for the construction phase of the house building. The life of this loan is typically 6 months to 2 years before converting to a conventional style mortgage

Country	Rural area, away from the city
Course	One horizontal row of bricks or cement block.
Covenant Restricted Property	Property controlled by covenants which must be abided.
Covenant	An agreement written in a deed or other document promising performance or nonperformance of certain acts, or stipulating certain uses or restriction son use of the property
Crawl Space	Typically shallow, it is unfinished space located either between the first floor of a house which has no basement or in the attic, immediately under the roof It is used for visual inspection and access to pipes, vents and / or ducts
Credit History	A report prepared by a specialized company detailing the history of an individuals debts and credit worthiness.
Crown Molding	A decorative material having a plane or curved narrow surface prepared for ornamental application at the wall and ceiling joints.
Designer Colors	Color upgrades beyond the typical selections that usually cost more.
Door Casings	Enclosed frame around the door
Door Sills	A horizontal piece of wood which forms the bottom portion of the door frame.
Door Levers	Lever device used to open doors (instead of a traditional door knob).
Down Spout	A vertical metallic tube that discharges the rain and melted snow from the gutter to the ground, or to a drainage system
Drywall	Also known as plaster board consisting of a gypsum core faced with heavy paper on both sides Used to cover interior walls and ceilings.
Duct	An enclosed rectangular or circular tube used to transfer hot and cold air to different parts of the house.
e-mail	Electronic communication memo sent / received within a linked computer system (such as the Internet)
Easement	A right or privilege that one party has in the property of another that entitles the holder to a specific limited use of the property.
Estimates	A rough evaluation or calculation of work to be done
Excavate	To dig out and remove earth to form a cavity for construction of a house or building.
Exterior Millwork	Building materials for the house exterior made of finished wood and manufactured in millwork plants fall under the term "millwork"
Fascia	The flat board enclosing the exterior ends of the rafters of the roof.
Felt	A heavy paper of organic or similar fibers impregnated with asphalt and used in building construction
Fill	Material (dirt, sand or gravel) used to raise the elevation of the ground.
Final Walk-Through	Final check of the finished house to ensure all is up to ones expectations and specifications
Finish Plumbing	Final phase of the plumbing including installing toilets, sinks, tubs and faucets.
Finish Heating & Cooling	Final phase of the heating and cooling including installing the vents and temperature controllers
Finish Electrical	Final phase of the electrical including installing the cover plates and finished light fixtures
Fireplace	Framed opening made in a chimney - or - a metal container with a smoke pipe used to hold an open fire.
Fixtures	Something that is attached as a permanent appendage or structural part of the house
Flashing	Sheet metal used in waterproofing roof grooves or angles.
Floor Joist	Framing pieces which rest on the outer foundation walls and interior beams or girders
Floor Drains	Pipe installed in the floor to permit drainage of water

Flooring	Floor base material.
Flue	A pipe that carries the smoke and exhaust gases of the fireplace upward to the outside atmosphere.
Footer	A concrete base poured over the bearing soil to distribute the load of the foundation walls above it over a wide area of soil
Foundation	The part of the structure below the first floor or below grade which safely supports the house throughout its life.
Foyer	An entrance hallway.
Framing Wood	Wood used in the construction of the skeleton of a house, to include walls, floors, ceiling and roof
Framing Materials	Studs, rafters, joists, sole plates and roof plates are put together to form the skeleton of the house
French Drain	Ground surrounding the house containing special holed piping then filled with rock and gravel to permit the exiting of water away from the house
Frost Line	The depth below the surface of the earth at which subsurface water may freeze during cold weather. This depth varies from regions and is set by the local building codes.
Gable Walls	Triangular part of a wall under the inverted "v" of the roof line.
General Contractor	A person who contracts to build a house or building, or a part of it, for another person for a profit.
GFI Circuits	Ground Fault Interrupter Circuits measure for electrical current leakage. An imbalance will cause the current to be immediately cut.
Girder	A large or main beam that supports numerous joists and heavy loads along its span.
Grade	The ground level or elevations Also, the slope of the surface of a lot
Graded	The slope of the surface or lot.
Great Room	Large room previously known as the "Family Room", used for watching television, relaxing, or other casual activities.
Grout	A mortar made of sand and cement to fill the joints between tiles.
Gutter Guards	Protective covers to allow rain and melted snow to enter the gutters but nothing else
Gutters	Lightweight metal or plastic troughs installed along the fascia to collect the rain and melted snow from the roof
Hand-Money	Money provided to the seller as a deposit to secure the property before the loan is approved
Hearth	The floor of the fireplace, typically made of brick, stone or cement
Heat Pump	An electric unit that cools the house during hot weather by absorbing heat from inside and discharging it to the outside It heats the house during cold weather by absorbing heat from outside and discharging it inside.
Hip	The external angle formed by the meeting of two sloping sides of a roof.
Home Shows	A gathering of many different businesses associated with any aspect of homes for the public. Businesses can range from the construction / financing stage, through furnishing and home repair stage.
Home Building Diary	Noteworthy accounts during the building process that you write into some type of notebook, calendar or logbook.
Homeowner's Insurance	Insurance for the homeowner to ensure replacement of any damage or liability coverage on the property
Homeowner's Association	An organization of homeowners which includes a Board of Directors to ensure the restrictive covenants are enforced
Hot Tips	Suggestions of things that the authors have learned from their own experience in bing Owner-Builders.
House Plan Layout	Drawings that indicates the details of construction of a house Typically included are the horizontal and vertical views (elevations) of the house.

Insulation	Material installed in the walls and ceiling to conserve energy.
Insurance Provider	Insurance company providing insurance.
Insurance Deductible	This is the pre-determined amount of pocket money required to be paid on a claim before the insurance coverage begins
Insured	A person whose property is insured
Interest Rate	The amount that the lenders charge borrowers for using their money.
Interior Trim	Material used to cover / trim the interior framed areas, walls and ceilings. Can include doors and countertops
Joiner	Equipment which joins pieces of wood together
Joint Treatment	Material used for joint treatment in gypsum wallboard finish
Joists	Horizontal beams that support a floor or ceiling
Laminate Style Tops	Thin plastic protective covering (laminate) adhered to a thick wood product for use in counter tops.
Land Survey	A map or plot made by a licensed surveyor showing the results of measuring the land with its elevations, improvements, boundaries and its relationship to surrounding tracts of land
Land Plot Plan	A map or chart of the lot, subdivision or community drawn by a surveyor showing boundary lines, buildings, improvements on the land and easements.
Landscaping	The shrubs, bushes and trees planted around the house in an organized manner designed to enhance the property.
Lath	Material (wood, metal, insulating board) that is fastened to the frame of a building to act as a plaster base
Lead Time	Amount of time notification necessary before a Sub-Contractor can start a job.
Lending Institution	Bank, Savings & Loan Association, Credit Union or other organization who will provide the financial support for the loan.
Level	Perfectly horizontal (level).
Liabilities	Debts or obligations of a person
Licensed	Someone who is authorized by a formal license to perform a specific speciality
Limestone (2B)	Limestone (approximately 1 to 1 ½ inch in diameter) rock used for drainage (such as French Drains) and driveway coverage.
Lintel	A horizontal architectural member spanning and usually carrying the load above on opening such as a door or window.
Loan Offerings	Conditions applied by the lending institution to obtain a loan (such as points, pre-payment items, certain insurance, etc...).
Local Municipality	A township, borough, city or village incorporated for local self-government.
Lock Pins	Small cylinder pins in a lock to prevent access without the appropriate key or opening device.
Loft	An upper room or floor
Lot	A piece of land that may be used as a building site for a house.
Low Voltage Switching	5 Volt DC Switching
Main Line	Main service water line in the house.
Mantel	The decorative framing around the fireplace.
Masonry Veneer	Walls built by a mason using a veneer (thin facing) of brick, stone, tile or similar materials that will cover a less expensive surface

Millwork	Building materials made of finished wood and manufactured in millwork plants fall under the term "millwork" It includes doors, window and door frames, mantels, stairways, moldings, but typically does not include flooring, ceiling or siding
Mine Subsidence Insurance	An insurance policy that protects the owner of real estate from any loss due to mines subsiding beneath the property.
Minimum Square Footage	Minimum amount of living space that a house must have to be built in certain restricted areas
Mining Report	Government report of the status of mines beneath the property It should include the depth of the mine from property surface and the last activity of the mine
Mirror Image	The house plan that has its parts reversely arranged in comparison with another
Model Home	Home constructed for purposes of showing or modeling the features for potential home buyers of what their home could look like.
Model Elevation	Drawings of the front and side views of a house.
Molding	A strip of decorative material prepared for ornamental application. These strips are often used to hide gaps at wall junctures
Money Dispersements	Payment distributed by having providing proof of the expense.
Monthly Pay-Back	Amount of monthly payment required on the loan
Morning Room	Similar to a Breakfast Nook Area where it is located near the kitchen area and typically used for breakfast.
Mortar	A mixture of cement, lime, sand and water to form a bonding material for brick laying, block laying or tile setting
Mortgage	The owner conditionally transfers the title of property to another as security for payment of a debt The owner retains possession and use of the land and once the debt is paid, the mortgage becomes void.
Note	A written acknowledgment of debt, such as a promissory note.
Occupancy Permit	A written warrant or license granted by one having authority allowing occupancy of the home or building, typically obtained once the construction is complete.
Occupancy	The act of taking or holding possession or becoming an occupant.
Offsets	To become marked by displacement.
Out-of-Pocket Expenses	Money spent from your budget; also, money spent beyond the Contractor Loan
Overruns	Expenses which have exceeded the budgeted amount per specific line items on the budget / cost sheet.
Owner-Builder	Someone who acts as their own General Contractor in building their own home.
Parcel	A lot or piece of land.
Partition Framing	Walls that subdivide spaces within any story of a building
Perk Test	For a septic system, a test conducted putting a hole in the ground and timing the absorption of a predetermined amount of water into the earth
Permit	A written warrant or license granted by one having authority
Pier	A vertical structural support, usually a masonry or metal column used to support the house, porch or deck.
Plan Drawings	The design drawings of a house showing all the elevations, plans and details needed to construct the house
Planer	Tool used for smoothing (planing) the wood surface.
Plot Plan	A drawing showing the top view of the plot of land.
Plumb	Perfectly vertical (plumb).

Portable Toilet Facility	Rented toilet facilities serviced on a frequent basis ensuring proper disposition of waste.
Posts	A vertical supporting member.
Pre-Hung Door	A doors which is purchased with frame and hinges attached. It may also have the casing installed
Programmable Light Switches	Lights timed (programmed) for on and off periods
Programmable Security Code	House security that requires a pre-programmed code entered to permit entrance.
Property Lines	Edges of the owned property
Property Age of Neighborhood	Average age of homes in the area.
Property Corners	Corners or intersecting edges of the property lines.
Property Values of Neighborhood	Average value of the homes in the area
Property	A piece of land, a lot
Quote	The bid offer of the service to be performed.
R-Value	A measure of the resistance of a building material to heat flow. Higher R-Values indicate better thermal insulating characteristics
Rafter	One of a series of structural roof members spanning from an exterior wall to a center ridge beam or ridge board.
Railing	The horizontal top or bottom bar extending over or between posts.
Rate Information Data Sheet	Sheet containing interest rates for loans and typically include some of the conditions for the loan.
Rebar	Steel rods placed in concrete to increase the strength of the concrete.
Receipt	A written acknowledgment of receiving goods or money.
Reimbursement	A payback of something after having shown proof of purchase with receipts.
Restrictive Covenant	An agreement written in a deed or other document promising performance or nonperformance of certain acts, or stipulating certain uses or restriction son use of the property.
Rise	The vertical distance between two points.
Rock Surface	Solid rock found beneath a layer of topsoil and earth.
Rosettes	An ornamental trim typically used for corners, joining trim or molding together.
Rough Electric	Installation of the electric wiring in the walls, floors or ceilings before the walls are covered with drywall, plaster, paneling or other wall covering.
Rough Heating & Cooling	Installation of heating and cooling in the walls, floors or ceilings before the walls are covered with drywall, plaster, paneling or other wall covering.
Rough Plumbing	Installation of plumbing in the walls before the walls are covered with drywall, plaster, paneling or other wall covering.
Run	The horizontal distance.
Sash	The frame in which a pane or panes of glass are set in a window; the movable part of a window.
Satellite Dish	TV reception dish for television channels beyond the local public accessible programs.
Sauna	A room for a dry heat steam bath.
Scaffolding	A system of temporary platforms for workers during the construction phase for use during things such as brick laying or painting
Sealer	A finishing material applied directly over uncoated wood for the purpose of sealing the surface

Security System	System put in place to provide notification if tampering has occurred Notification can include sending a signal to the local police.
Septic Systems	Contains a settling tank in which the sludge in the household sewage settles and the effluent discharges into an absorption field or seepage pit
Service Door	Door used to exit the garage to the outside without having to open the main garage door.
Set Back	The minimum distance between the street and the building line It is established by the local zoning ordinances or deed restrictions.
Shaper / Router	Tool used to shape wood into having a detailed finished edging.
Sheathing	The first covering of boards or material on the outside wall or roof prior to installing the finished siding or roof covering
Shell Framing	The house shell frame. This is the initial part of the house construction.
Shingles	Pieces of wood or other material used as an overlapping outer covering on walls or roofs.
Shutter	Lightweight wood or non-wood frames in the form of doors located at each side of a window
Side Entry Garage	Garage entrance on one of the sides of the house
Siding	Finish covering of the outside wall of a frame building typically of horizontal boards of plastic, aluminum or other light-weight material
Sill Plate	The horizontal member forming the base of a window or foot of a door. Also, the lowest member of the house framing resting on top of the foundation wall.
Site Preparations	Can include clearing the land of unwanted trees and weeds in preparation for the excavation to begin.
Skylight	An opening in the roof covered with thick glass allowing light into the area below.
Slab	A thick plate made of concrete placed directly on earth or a gravel base and usually about four inches thick.
Soffit	The boards covering the underside of the eave overhang
Springs	Sources of water issuing from the ground.
Square	Forming a right angle.
Stain	A die used for finishing wood surfaces
Stakeout	Pieces of wood inserted in the ground at the corners and boundary lines of a piece of property to precisely define its boundaries
Stanchion Post	An upright post to support the roof
Stories	Levels or floors in a house or building
Street Accessibility	How close the house will be to the street and if there are properties that must be traveled through to get to the street.
Street Map	A map provided by the Local Municipality containing all the local streets in the neighborhood
Stucco	A fine plaster used for decoration and ornamentation of interior walls
Studs	The vertical members in wall framing. Their function is to form the walls and support the floors and the roof above them. They are typically placed either 16 inches or 24 inches apart.
Sub-Contractor	A professional who specializes in one or more disciplines of construction such as excavation, plumbing, etc He / She may be hired to provide labor only or labor and materials.
Sub-Floor	Plywood or boards nailed directly to the floor joists to form a base for the finish flooring
Sub-Lines	Smaller diameter water lines off the main in-house water line
Suburbs	Smaller, outlying community adjacent to a city or town.

Sun Room	A glass enclosed porch or living room with a sunny exposure
Survey	A drawing made to scale showing the lengths and directions of the boundary lines of the lot, the surrounding lots and streets, the position of the house and all exterior items (driveway, walkway, deck) and any existing encroachments.
Tap-In Fees	Fees paid to "tap-in" to public utilities such as water, sewage, electrical and gas
Tax Milage	Number of mils (thousandths or 10^{-3}) of taxes required to pay on the assessed (or portion of the assessed) property value.
Taxes	A charge imposed by authority on property for public purposes.
Temporary Utilities	Utilities such as electric and water set up during the initial construction phase used on a temporary basis until the permanent connections are set up
Term	The length of time by the end of which the mortgage loan must be paid in full.
Terrace	An open platform or porch.
Tongue and Groove	Carpentry joint in which the jutting edge of one board fits into the grooved end of a similar board
Topsoil	Surface of soil including the organic layer in which plants have their most roots
Total Wall Thickness	The thickness of the wall which includes the framing materials, wall or drywall and any decorative surface The insulation material typically fits within the framed wall.
Township Ordinances	The local community regulations or laws set forth which must be abided by to live in the neighborhood
Trash Haulers	Company which will remove the excessive construction trash generated during the building phase A large dumpster is typically provided for use on a rental basis
Trash Bins	Bins or dumpsters used for containing the construction trash generated during the building phase.
Trash Generators	Excessive amounts of trash are generated during the many phases of construction including framing, brick laying, electrical, insulation, drywall, and trim phases of construction
Tray Style Ceiling	Decorative, raised, flat ceiling providing a layered look.
Trim	Finish carpentry, including installation of interior doors, window and door casings, closets, shelves, molding and hardware
Trim Carpenter	A worker who installs the trim phase of home construction
Truss	A combination of structural members usually arranged in triangular units to form a rigid framework for spanning between load bearing walls.
Underlayment	Any paper or felt composition used to separate the roofing deck from the shingles providing a smooth, even surface for applying the finish
Urban / City	An inhabited place of greater size and population.
Utilities	Services such as light, power or water, provided by a public service
Utility Companies	Company providing utility service to the community, typically government regulated.
Valley	The internal angle formed by the junction of two sloping sides of a roof
Vapor Barrier	Material, such as paper, plastic, metal or paint, which is used to prevent vapor from passing from rooms into the outside walls.
Walk-Off Items	Construction items small enough that can be easily picked up by someone walking by and taken from the construction site.
Wall Covering	Materials such as paint, wall paper, or paneling to cover the primed wall.
Washtub	A tub located in the laundry facility in which clothes are washed or soaked.
Waterproofing	Making a foundation impervious to water

Weatherstripping	Thin metal or other material to prevent infiltration of air and moisture around windows and doors.
Whole House Systems	Systems such as intercoms, vacuums, or music that are dispersed and available throughout the whole house
Window Sills	A horizontal piece of wood which forms the bottom portion of the window frame.
Window Casings	Enclosed frame around the window.
Wood Species	Type of wood (such as cherry, oak, maple, pine, etc. .)
Zoning Classification	Local regulations affecting property uses and the type of construction (such as residential or commercial).

Index

W

Z

Order Form

Build Your Own Home from Start to Finish
Go Through a Step-by-Step Process of:
- **Determining House Specs**
- **Finding an Owner / Builder Program**
- **Selecting a Lot**
- **Getting a Contractor Loan**
- **Selecting Sub-Contractors**
- **Building Your House**

And Save as much as 25% of the Total Construction Costs

by Dennis and Nancy Sue Swoger

ISBN 0-9626692-9-6

Price: $24.95
Sales Tax: Please add 6% for books shipped to Pennsylvania addresses. Add an additional 1% for books shipped to the Philadelphia, PA area or Allegheny County, PA.

Please send Check or Money Order.

Mail Orders to:

QP Publishing
PO Box 237
Finleyville, Pennsylvania 15332-0237 USA

Please send the book **"Build Your Own Home from Start to Finish"** to:

Name: _____

Address: _____

City: _____ **State:** _____ **Zip:** _____

Telephone: (_____)_____

Thank-You!